Math Challenge II-B
Geometry

Areteem Institute

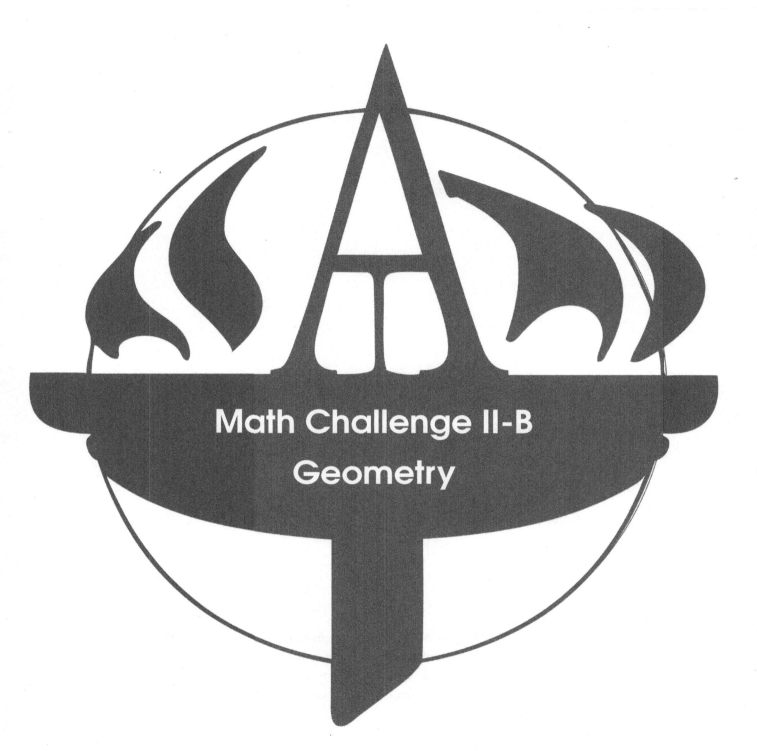

Math Challenge II-B Geometry

Edited by David Reynoso
John Lensmire
Kevin Wang
Kelly Ren

WWW.ARETEEM.ORG

PUBLISHED BY ARETEEM PRESS

ISBN: 1-944863-19-2
ISBN-13: 978-1-944863-19-7
First printing, August 2018.

TITLES PUBLISHED BY ARETEEM PRESS

Cracking the High School Math Competitions (and Solutions Manual) - Covering AMC 10 & 12, ARML, and ZIML
Mathematical Wisdom in Everyday Life (and Solutions Manual) - From Common Core to Math Competitions
Geometry Problem Solving for Middle School (and Solutions Manual) - From Common Core to Math Competitions
Fun Math Problem Solving For Elementary School (and Solutions Manual)
Fun Math Problem Solving For Elementary School Vol. 2

ZIML MATH COMPETITION BOOK SERIES

ZIML Math Competition Book Division E 2016-2017
ZIML Math Competition Book Division M 2016-2017
ZIML Math Competition Book Division H 2016-2017
ZIML Math Competition Book Jr Varsity 2016-2017
ZIML Math Competition Book Varsity Division 2016-2017
ZIML Math Competition Book Division E 2017-2018
ZIML Math Competition Book Division M 2017-2018
ZIML Math Competition Book Division H 2017-2018
ZIML Math Competition Book Jr Varsity 2017-2018
ZIML Math Competition Book Varsity Division 2017-2018
ZIML Math Competition Book Division E 2018-2019
ZIML Math Competition Book Division M 2018-2019
ZIML Math Competition Book Division H 2018-2019
ZIML Math Competition Book Jr Varsity 2018-2019
ZIML Math Competition Book Varsity Division 2018-2019

MATH CHALLENGE CURRICULUM TEXTBOOKS SERIES

Math Challenge I-A Pre-Algebra and Word Problems
Math Challenge I-B Pre-Algebra and Word Problems
Math Challenge I-C Algebra
Math Challenge II-A Algebra
Math Challenge II-B Algebra
Math Challenge III Algebra
Math Challenge I-A Geometry
Math Challenge I-B Geometry
Math Challenge I-C Topics in Algebra

Math Challenge II-A Geometry
Math Challenge II-B Geometry
Math Challenge III Geometry
Math Challenge I-A Counting and Probability
Math Challenge I-B Counting and Probability
Math Challenge I-C Geometry
Math Challenge II-A Combinatorics
Math Challenge II-B Combinatorics
Math Challenge III Combinatorics
Math Challenge I-A Number Theory
Math Challenge I-B Number Theory
Math Challenge I-C Finite Math
Math Challenge II-A Number Theory
Math Challenge II-B Number Theory
Math Challenge III Number Theory

COMING SOON FROM ARETEEM PRESS

Fun Math Problem Solving For Elementary School Vol. 2 Solutions Manual
Counting & Probability for Middle School (and Solutions Manual) - From Common Core to Math Competitions
Number Theory Problem Solving for Middle School (and Solutions Manual) - From Common Core to Math Competitions

The books are available in paperback and eBook formats (including Kindle and other formats).
To order the books, visit `https://areteem.org/bookstore`.

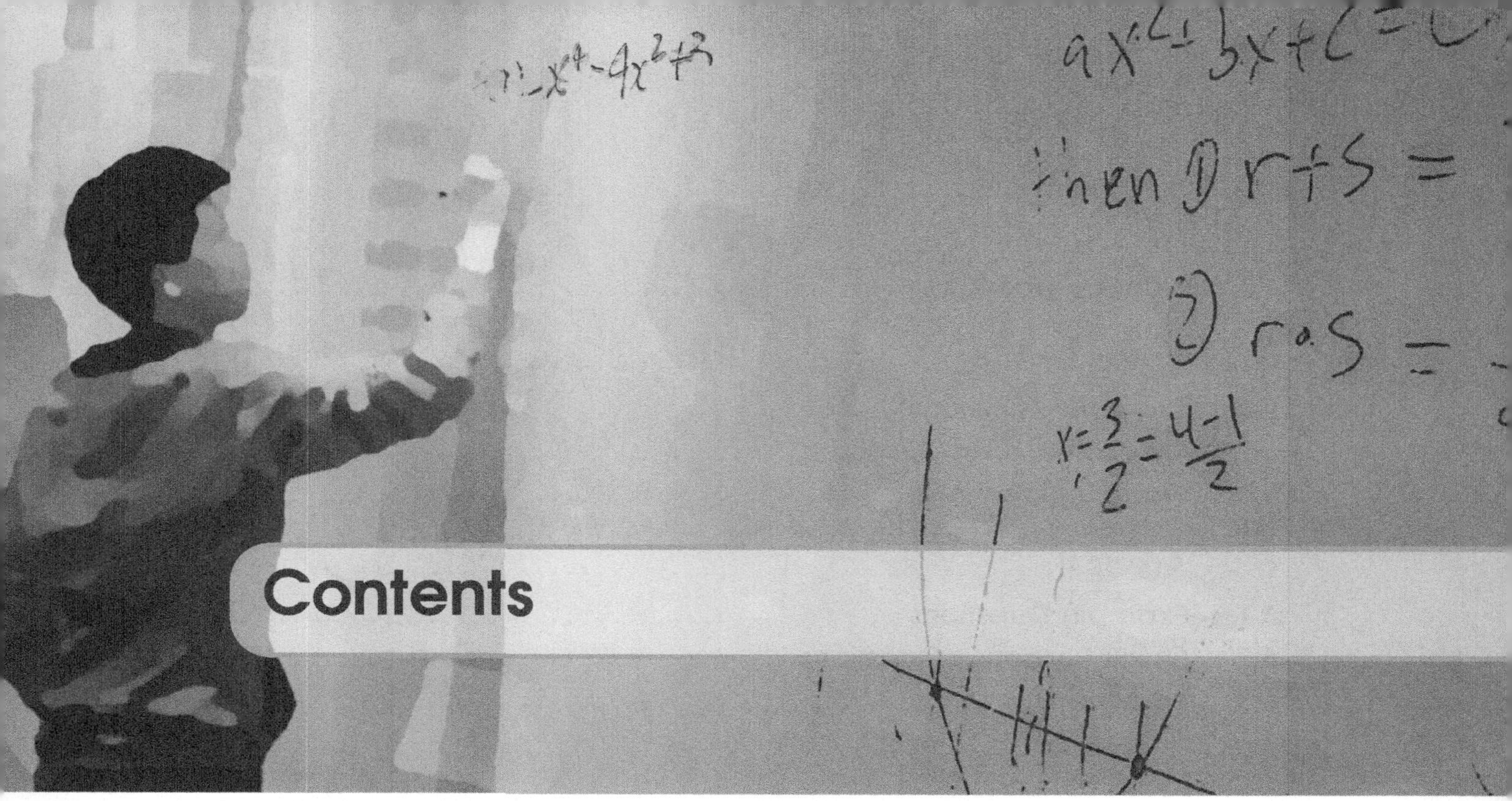

Contents

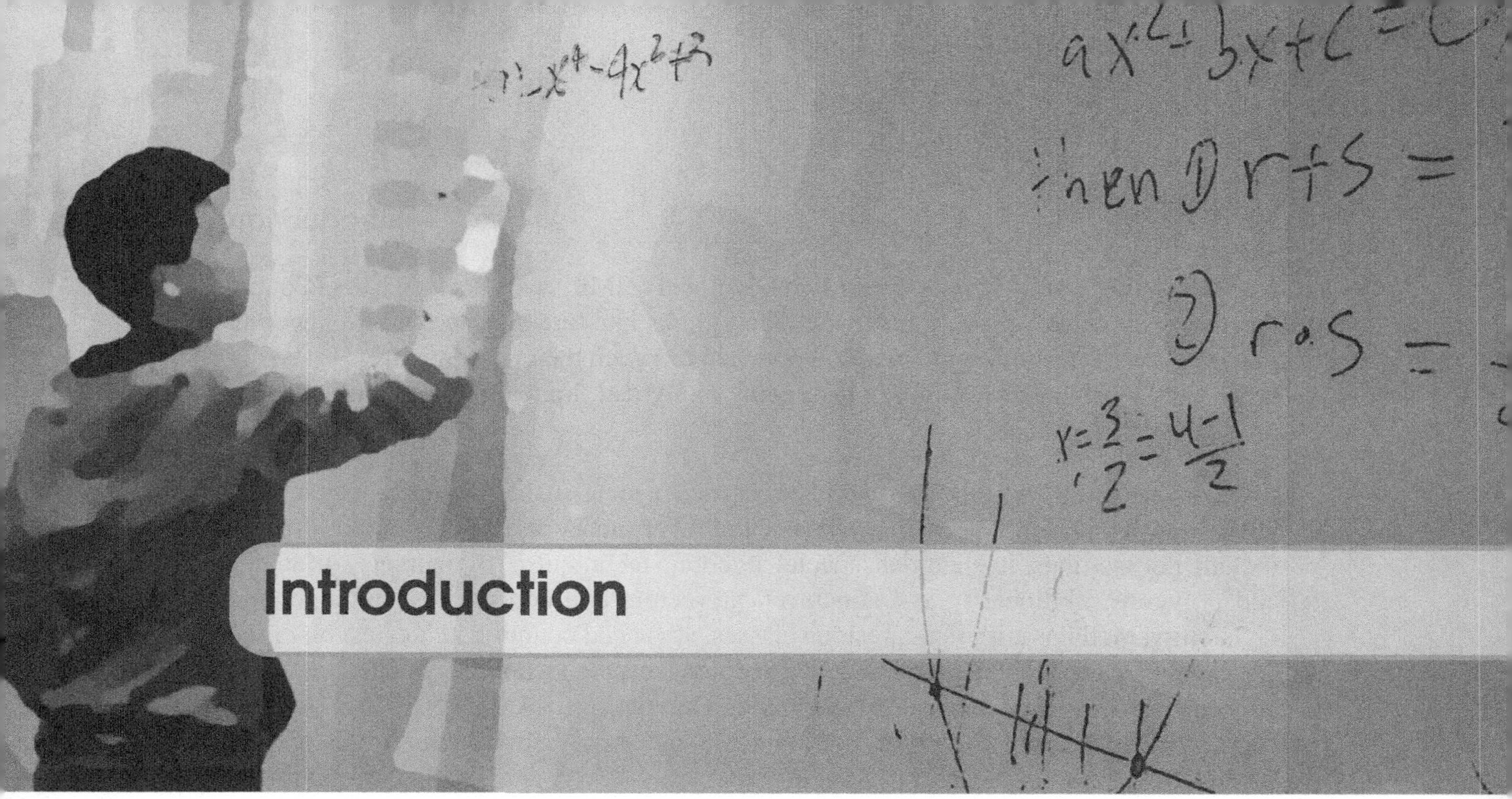

Introduction

The math challenge curriculum textbook series is designed to help students learn the fundamental mathematical concepts and practice their in-depth problem solving skills with selected exercise problems. Ideally, these textbooks are used together with Areteem Institute's corresponding courses, either taken as live classes or as self-paced classes. According to the experience levels of the students in mathematics, the following courses are offered:

- Fun Math Problem Solving for Elementary School (grades 3-5)
- Algebra Readiness (grade 5; preparing for middle school)
- Math Challenge I-A Series (grades 6-8; intro to problem solving)
- Math Challenge I-B Series (grades 6-8; intro to math contests e.g. AMC 8, ZIML Div M)
- Math Challenge I-C Series (grades 6-8; topics bridging middle and high schools)
- Math Challenge II-A Series (grades 9+ or younger students preparing for AMC 10)
- Math Challenge II-B Series (grades 9+ or younger students preparing for AMC 12)
- Math Challenge III Series (preparing for AIME, ZIML Varsity, or equivalent contests)
- Math Challenge IV Series (Math Olympiad level problem solving)

These courses are designed and developed by educational experts and industry professionals to bring real world applications into the STEM education. These programs are ideal for students who wish to win in Math Competitions (AMC, AIME, USAMO, IMO,

ARML, MathCounts, Math League, Math Olympiad, ZIML, etc.), Science Fairs (County Science Fairs, State Science Fairs, national programs like Intel Science and Engineering Fair, etc.) and Science Olympiad, or purely want to enrich their academic lives by taking more challenges and developing outstanding analytical, logical thinking and creative problem solving skills.

In Math Challenge II-B, students learn and practice in areas such as algebra and geometry at the high school level, as well as advanced number theory and combinatorics. Topics include polynomials, inequalities, special algebraic techniques, trigonometry, triangles and polygons, collinearity and concurrency, vectors and coordinates, numbers and divisibility, modular arithmetic, residue classes, advanced counting strategies, binomial coefficients, and various other topics and problem solving techniques involved in math contests such as the American Mathematics Competition (AMC) 10 & 12, ARML, beginning AIME, and Zoom International Math League (ZIML) Junior Varsity and Varsity Divisions.

The course is divided into four terms:

- Summer, covering Algebra
- Fall, covering Geometry
- Winter, covering Combinatorics
- Spring, covering Number Theory

The book contains course materials for Math Challenge II-B: Geometry.

We recommend that students take all four terms. Each of the individual terms is self-contained and does not depend on other terms, so they do not need to be taken in order, and students can take single terms if they want to focus on specific topics.

Students can sign up for the course at `classes.areteem.org` for the live online version or at `edurila.com` for the self-paced version.

About Areteem Institute

Areteem Institute is an educational institution that develops and provides in-depth and advanced math and science programs for K-12 (Elementary School, Middle School, and High School) students and teachers. Areteem programs are accredited supplementary programs by the Western Association of Schools and Colleges (WASC). Students may attend the Areteem Institute in one or more of the following options:

- Live and real-time face-to-face online classes with audio, video, interactive online whiteboard, and text chatting capabilities;
- Self-paced classes by watching the recordings of the live classes;
- Short video courses for trending math, science, technology, engineering, English, and social studies topics;
- Summer Intensive Camps held on prestigious university campuses and Winter Boot Camps;
- Practice with selected free daily problems and monthly ZIML competitions at ziml.areteem.org.

Areteem courses are designed and developed by educational experts and industry professionals to bring real world applications into STEM education. The programs are ideal for students who wish to build their mathematical strength in order to excel academically and eventually win in Math Competitions (AMC, AIME, USAMO, IMO, ARML, MathCounts, Math Olympiad, ZIML, and other math leagues and tournaments, etc.), Science Fairs (County Science Fairs, State Science Fairs, national programs like Intel Science and Engineering Fair, etc.) and Science Olympiads, or for students who purely want to enrich their academic lives by taking more challenging courses and developing outstanding analytical, logical, and creative problem solving skills.

Since 2004 Areteem Institute has been teaching with methodology that is highly promoted by the new Common Core State Standards: stressing the conceptual level understanding of the math concepts, problem solving techniques, and solving problems with real world applications. With the guidance from experienced and passionate professors, students are motivated to explore concepts deeper by identifying an interesting problem, researching it, analyzing it, and using a critical thinking approach to come up with multiple solutions.

Thousands of math students who have been trained at Areteem have achieved top honors and earned top awards in major national and international math competitions, including Gold Medalists in the International Math Olympiad (IMO), top winners and qualifiers at the USA Math Olympiad (USAMO/JMO) and AIME, top winners at the

Zoom International Math League (ZIML), and top winners at the MathCounts National Competition. Many Areteem Alumni have graduated from high school and gone on to enter their dream colleges such as MIT, Cal Tech, Harvard, Stanford, Yale, Princeton, U Penn, Harvey Mudd College, UC Berkeley, or UCLA. Those who have graduated from colleges are now playing important roles in their fields of endeavor.

Further information about Areteem Institute, as well as updates and errata of this book, can be found online at `http://www.areteem.org`.

Acknowledgments

This book contains many years of collaborative work by the staff of Areteem Institute. This book could not have existed without their efforts. Huge thanks go to the Areteem staff for their contributions!

The examples and problems in this book were either created by the Areteem staff or adapted from various sources, including other books and online resources. Especially, some good problems from previous math competitions and contests such as AMC, AIME, ARML, MATHCOUNTS, and ZIML are chosen as examples to illustrate concepts or problem-solving techniques. The original resources are credited whenever possible. However, it is not practical to list all such resources. We extend our gratitude to the original authors of all these resources.

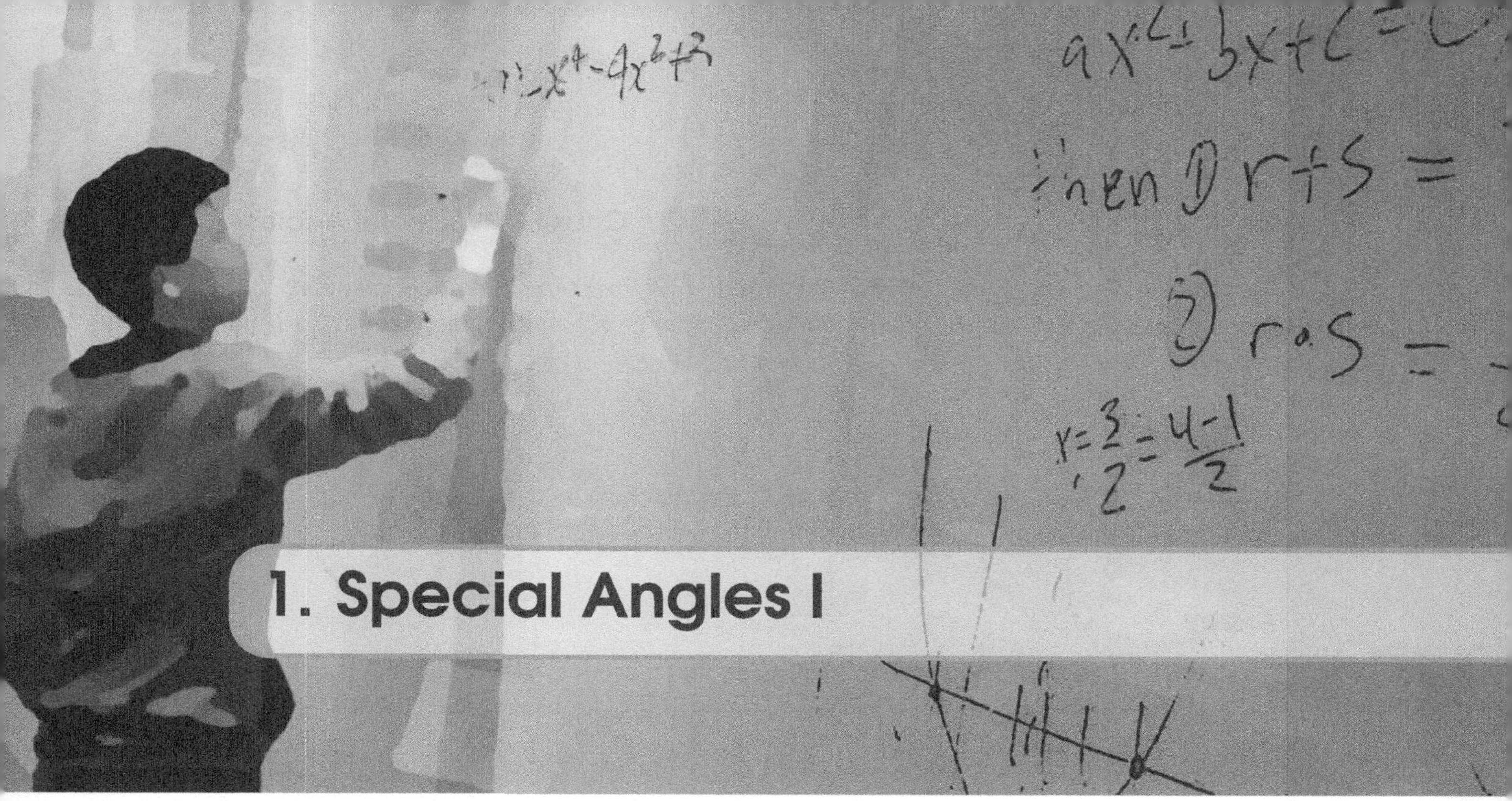

1. Special Angles I

Special Angles

- Special angles are the angles 30°, 60°, 90° and 45°.
- An equilateral triangle has three 60° angles.
- In $\triangle ABC$, if $\angle C = 90^\circ$, $\angle A = \angle B = 45^\circ$, then $AB = \sqrt{2}AC = \sqrt{2}BC$.
- In $\triangle ABC$, if $\angle C = 90^\circ$, $\angle A = 60^\circ$, and $\angle B = 30^\circ$, then $AB = 2AC$, $BC = \sqrt{3}AC$.

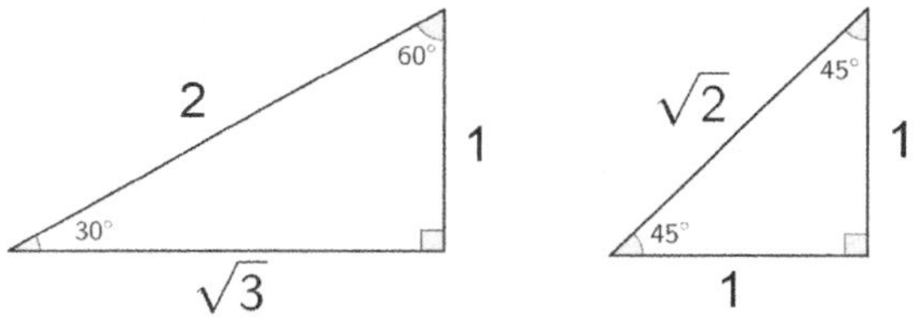

1.1 Example Questions

Problem 1.1 Suppose that $ABCD$ is a square, and that CDP is an equilateral triangle, with P *outside* the square. What is the size of angle PAD?

Problem 1.2 Given square $ABCD$, let P and Q be the points outside the square that make triangles CDP and BCQ equilateral. Segments AQ and BP intersect at T. Find angle ATP.

Problem 1.3 Squares $OPAL$ and $KEPT$ are attached to the outside of equilateral triangle PEA. Draw segment TO, then find the size of angle TOP.

Problem 1.4 Mark P inside square $ABCD$, so that triangle ABP is equilateral. Let Q be the intersection of BP with diagonal AC. Triangle CPQ looks isosceles. Is this actually true?

Problem 1.5 Let triangle ABC be equilateral triangle with side length 16. Let D be on side $\overline{AB}$ and E be on side $\overline{AC}$ such that $\overline{DE} \| \overline{BC}$. Assume triangle ADE and trapezoid $DECB$ have the same perimeter. What is the length of $\overline{AD}$?

Problem 1.6 Let E be a point inside unit square $ABCD$ such that CDE is an equilateral triangle. Find the area of triangle AEC.

Problem 1.7 A triangle has a 60-degree angle and a 45-degree angle, and the side opposite the 45-degree angle has length 12. How long is the side opposite the 60-degree angle?

Problem 1.8 Let ABC be a triangle with $AB = AC$ and $\angle BAC = 20^\circ$, and let P be a point on side AB such that $AP = BC$. Construct point D such that triangle ACD is equilateral, as shown in the diagram below. Show that triangle DCP is isosceles.

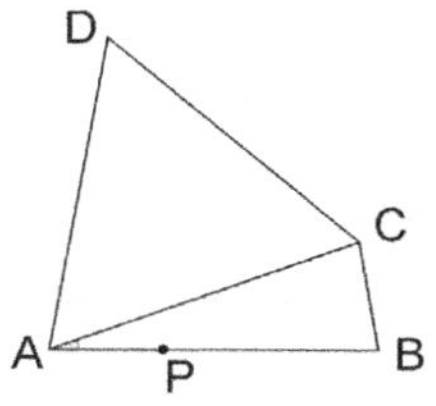

Problem 1.9 In triangle ABC, $AB = AC$, and BD is the altitude on AC. Given that $BD = \sqrt{3}$, and $\angle DBC = 60^\circ$, find the area of $\triangle ABC$.

Problem 1.10 In a right triangle $\triangle ABC$ suppose $\angle B = 90^\circ$ and $\angle C = 30^\circ$. Suppose point D is on $\overline{BC}$ with $\angle ADB = 45^\circ$ and $DC = 10$. Find the length of AB.

1.2 Quick Response Questions

Problem 1.11 If one of the leg lengths of a $45-45-90$ triangle is 3, the length of the hypotenuse is $A\sqrt{B}$. What is $A+B$?

Problem 1.12 If the shorter leg length of a $30-60-90$ triangle is 3, what is the length of the hypotenuse? Round your answer to the nearest tenth if necessary.

Problem 1.13 Consider the diagram below.

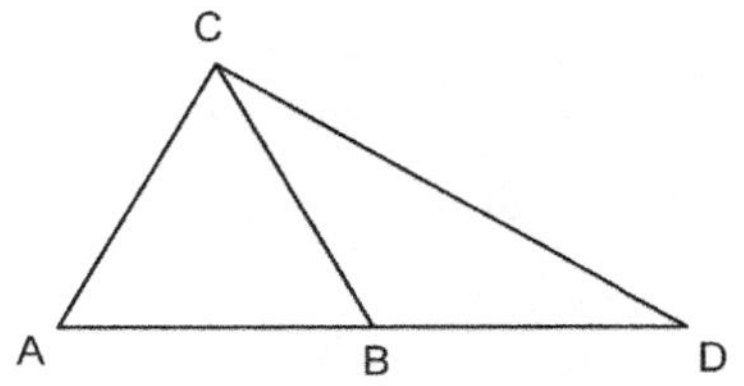

Suppose $AB = AC = 1$ and $\angle BAC = 60^\circ$ and $\angle ADC = 30^\circ$. Find BD.

Problem 1.14 Suppose an isosceles triangle has an angle of 30°. What is the measure of the largest angle the triangle may have?

Problem 1.15 Suppose you have an equilateral triangle. What are the angles of a right triangle created by drawing an altitude?

(A) $45-45-90$
(B) $30-60-90$
(C) $60-60-60$
(D) None of the above

Problem 1.16 Suppose you have an equilateral triangle and you form a right triangle by drawing an altitude. On the right triangle, how does the short side compare with the other two sides?

(A) The short side is half the hypotenuse, and $1/\sqrt{3}$ of the other leg.
(B) The short side is $1/\sqrt{3}$ of the hypotenuse and half of the other leg.
(C) They are all the same size.
(D) None of the above

Problem 1.17 The sides of an equilateral triangle are 4 cm long. An altitude of this triangle is $A\sqrt{B}$, what is $A+B$?

Problem 1.18 In the following diagram $\angle ADB = 90°$, $AD = 3$, $BD = 3$ and $BC = 6$. What is the measure of $\angle ABC$?

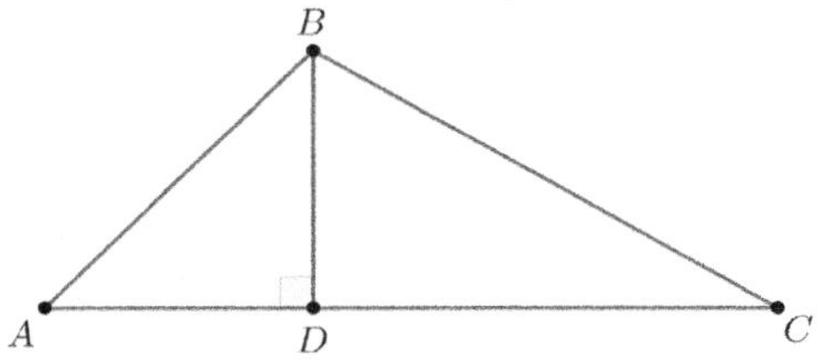

Problem 1.19 Suppose you have a trapezoid $ABCD$, with AB parallel to CD, $AD = 5\sqrt{2}$, $BC = 5\sqrt{2}$, $AB = 20$ and $CD = 10$. What is the area of the trapezoid?

Problem 1.20 On the following diagram $BD = 3$ and $AC = P + \sqrt{Q}$. What is $P + Q$?

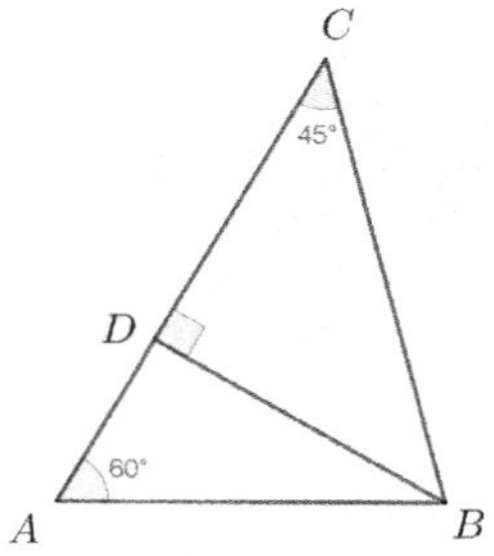

1.3 Practice Questions

Problem 1.21 Suppose that $ABCD$ is a square, and that CDP is an equilateral triangle, with P *inside* the square.

(a) What is the size of $\angle PAD$?

(b) What is the size of $\angle BPA$?

Problem 1.22 Given square $ABCD$, let P and Q be the points outside the square that make triangles CDP and BCQ equilateral. Show triangle APQ is equilateral.

Problem 1.23 Squares $OPAL$ and $KEPT$ are attached to the outside of equilateral triangle PEA. Decide whether segments EO and AK have the same length, and give your reasons.

Problem 1.24 Mark P inside square $ABCD$, so that triangle ABP is equilateral. Find the size of $\angle CPD$.

Problem 1.25 Let triangle ABC be equilateral triangle. Let D be on side $\overline{AB}$ and E be on side $\overline{AC}$ such that $\overline{DE} \| \overline{BC}$. Let the perimeter of triangle ADE to be 12 and the perimeter of trapezoid $DECB$ to be 16. Find the perimeter of triangle ABC.

Problem 1.26 Let E be a point outside unit square $ABCD$ such that CDE is an equilateral triangle. Find the area of triangle AEC.

Problem 1.27 A triangle has a 45-degree angle and a 30-degree angle, and the side opposite the 45-degree angle has length 12. How long is the side opposite the 30-degree angle?

Problem 1.28 Let ABC be a triangle with $AB = AC$ and $\angle BAC = 20^\circ$, and let P be a point on side AB such that $AP = BC$. Find $\angle ACP$.

Problem 1.29 Given that one of the angles of the triangle with sides $(5,7,8)$ is 60°, show that one of the angles of the triangle with sides $(3,5,7)$ is 120°.

Problem 1.30 In a right triangle $\triangle ABC$ suppose $\angle B = 90^\circ$ and $\angle A = 45^\circ$. Suppose point D is on $\overline{BC}$ with $\angle ADB = 60^\circ$ and $DC = 5$. Find the length of AB.

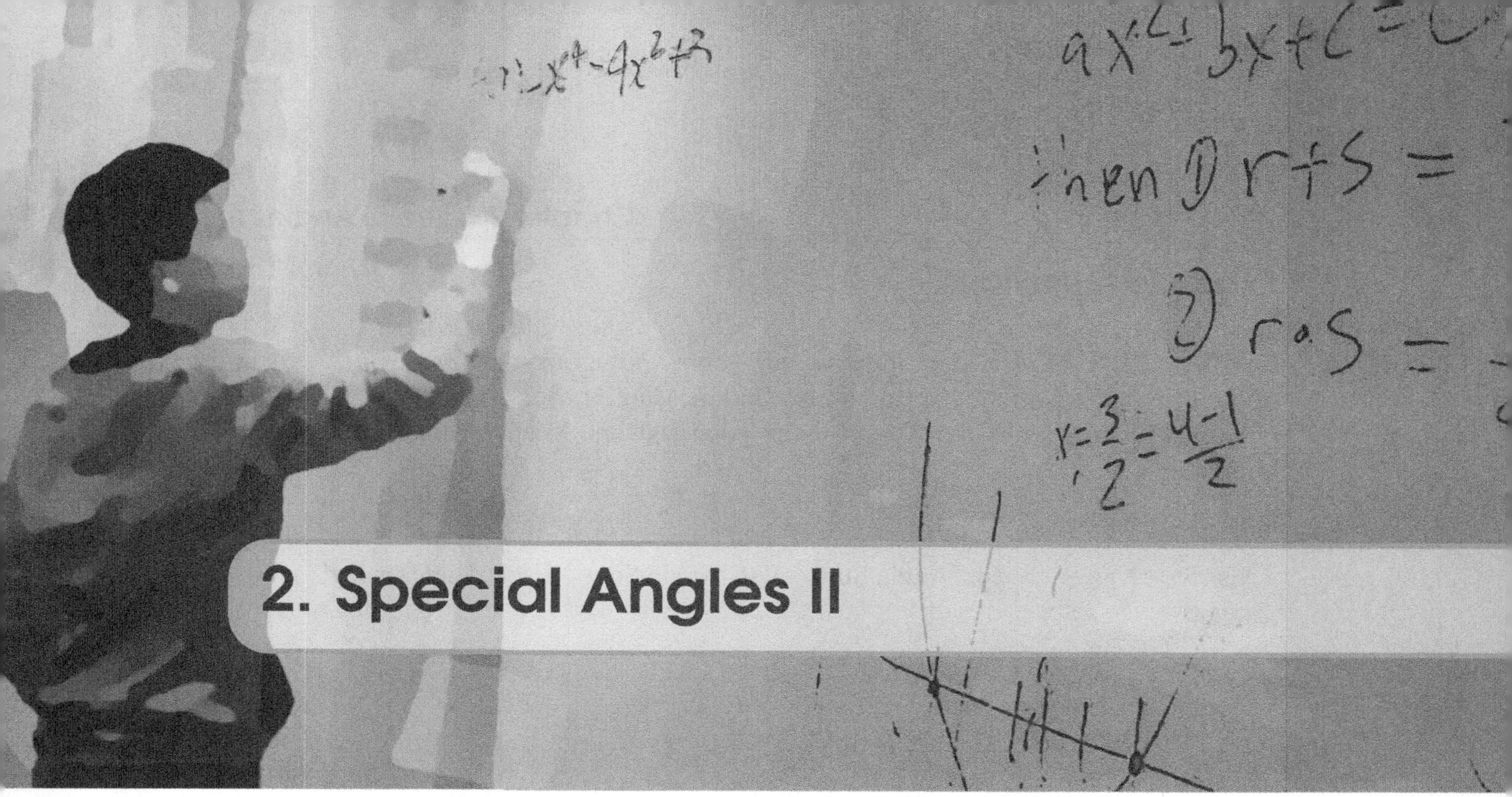

2. Special Angles II

Review and Definitions

- **Equilateral polygon:** A polygon that has all of its sides of the same length is called equilateral.
- **Equiangular polygon:** A polygon that has all of its internal angles the same size is called equiangular.
- **Regular polygon:** A polygon that is both equilateral and equiangular is called regular.
- We use $[ABC]$ to denote the area of triangle ABC, $[DEFG]$ to denote the area of quadrilateral $DEFG$, etc.

2.1 Example Questions

Problem 2.1 Complete the following table about polygons with n sides: name, sum of interior angles, sum of exterior angles, and measure of each angle in case of regular polygon. All angles are in degrees. Justify your answers. Keep the chart for your own reference.

n	Name	Int. Angle Sum	Ext. Angle Sum	Each Angle (if regular)
3	Triangle			
4				
5				
6				
7	Heptagon			
8				
9	Nonagon			
10				
12	Dodecagon			
20	Icosagon			

Problem 2.2 Use 6 equilateral triangles to form a hexagon $ABCDEF$.

(a) Show hexagon $ABCDEF$ is regular. Justify your answer.

(b) Calculate the angle AED.

Problem 2.3 Four non-overlapping regular plane polygons all have sides of length 1. The polygons meet at a point A in such a way that the sum of the four interior angles at A is $360°$. Among the four polygons, two are squares and one is a triangle. What is the last polygon?

Problem 2.4 Find the area of the largest equilateral triangle that fits in a regular hexagon of area 50.

Problem 2.5 In equiangular octagon $ABCDEFGH$, $AB = CD = EF = GH = 6\sqrt{2}$ and $BC = DE = FG = HA$. Given the area of the octagon is 184, compute the length of side BC.

Problem 2.6 Answer the following

(a) Given a 60-120 isosceles trapezoid, prove that the sum of the length of the top and one of the side is equal to the length of the base.

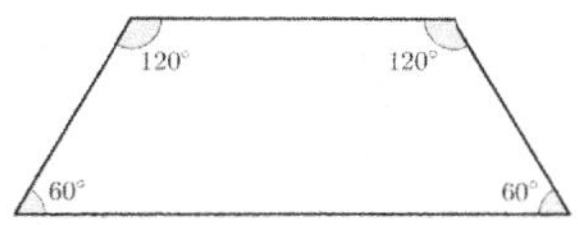

(b) Let triangle ABC be a equilateral triangle with length equals to 1. Let point P be the center of the triangle ABC, D be point on $\overline{BC}$, E be point on $\overline{CA}$, and F be point on $\overline{AB}$, so that $\overline{PD} \| \overline{AB}$, $\overline{PE} \| \overline{BC}$, and $\overline{PF} \| \overline{AC}$. Find the value of $PD + PE + PF$.

Problem 2.7 Show that a regular dodecagon (12-sided polygon) can be cut into pieces that are all regular polygons, which need not all have the same number of sides.

Problem 2.8 What is the side length of the largest equilateral triangle that can fit inside a 2-by-2 square?

Problem 2.9 Let $ABCD$ be a unit square and let P, Q be on sides $\overline{AD}, \overline{AB}$ respectively such that $\triangle APQ$ has perimeter 2. Rotate $\triangle PDC$ 90° about C. Call the point P is rotated to P'. Prove that $\triangle PQC$ is congruent to $\triangle P'QC$.

Problem 2.10 In convex quadrilateral $ABCD$, $\angle A = 60^\circ$, $\angle C = 30^\circ$, and $AB = AD$. Show that $AC^2 = BC^2 + CD^2$.

2.2 Quick Response Questions

Problem 2.11 Is the following a regular polygon?

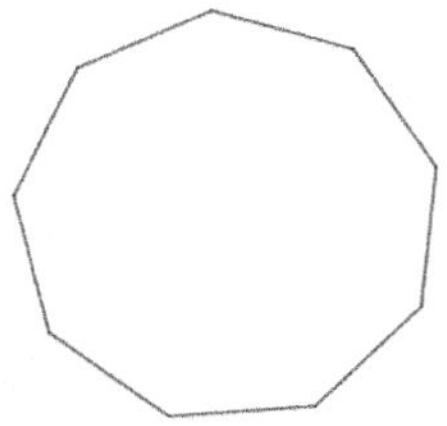

Problem 2.12 Is the following an equiangular polygon that is not equilateral?

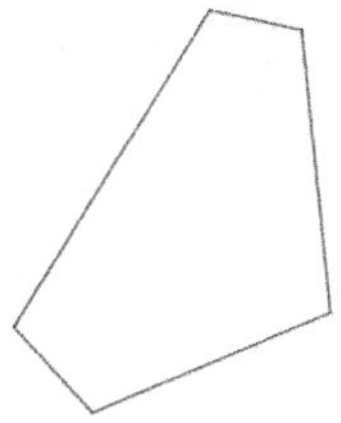

Problem 2.13 Is the following an equilateral polygon that is not equiangular?

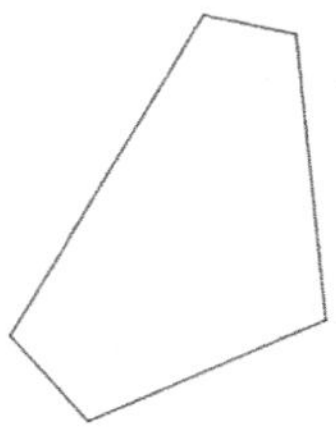

Problem 2.14 Let $ABCDEF$ be a regular hexagon, and let $EFGHI$ be a regular pentagon. $\angle GAF$ could have measure α or β degrees, where $\alpha > \beta$. Find $\alpha - \beta$.

Problem 2.15 A regular n-sided polygon has exterior angles of m degrees each. What is m?

(A) $m = 360$
(B) $m = 180(n-2)$
(C) $m = \frac{360}{n}$
(D) $m = 360(n-2)$

Problem 2.16 Let $\triangle ABC$ be an isosceles triangle with $\angle A = 50^\circ$. What is the sum of all possible angle measures of $\angle B$?

Problem 2.17 A regular n-sided polygon has exterior angles of m degrees each. For how many polygons is m a whole number?

Problem 2.18 Consider a regular nonagon $COMPUTERS$. Is it true that $TE + ES = SP$?

Problem 2.19 As shown in the diagram below, the equiangular convex hexagon $ABCDEF$ has $AB = 1$, $BC = 4$, $CD = 2$, and $DE = 4$. $[ABCDEF] = \frac{P\sqrt{Q}}{R}$, with $\gcd(P, R) = 1$ and such that R has no square factors. What is $P + Q + R$?

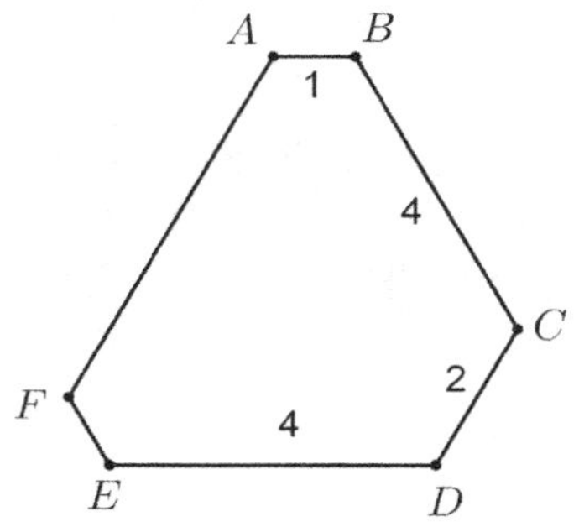

Problem 2.20 Find the area of the largest regular hexagon that fits in an equilateral triangle of area 108.

2.3 Practice Questions

Problem 2.21 Give two more examples of n so that a regular polygon with n sides has integer interior angles.

Problem 2.22 Use 6 equilateral triangles to form a hexagon $ABCDEF$, then find the area of triangle AED in terms of the total area.

Problem 2.23 Four non-overlapping regular plane polygons all have sides of length 1. The polygons meet at a point A in such a way that the sum of the four interior angles at A is $360°$. Among the four polygons, two are squares and one is a triangle. What is the perimeter of the entire shape?

Problem 2.24 Construct the largest equilateral triangle in a regular hexagon, and find the perimeter of the triangle given that the side lengths of the hexagon are 4.

Problem 2.25 In equiangular octagon $ABCDEFGH$, $AB = CD = EF = GH$ and $BC = DE = FG = HA$. Argue that $\overline{AB} \perp \overline{CD}$.

Problem 2.26 Let triangle ABC be a equilateral triangle with length equals to 1. Let point P be an arbitrary point in the interior of triangle ABC, D be point on $\overline{BC}$, E be point on $\overline{CA}$, and F be point on $\overline{AB}$, so that $\overline{PD} || \overline{AB}$, $\overline{PE} || \overline{BC}$, and $\overline{PF} || \overline{AC}$. Find the value of $PD + PE + PF$.

Problem 2.27 Come up with a pattern to cover the entire plane (called a *tiling*) with regular hexagons, squares, and triangles (each with the same side length). Do this in a way such that no two hexagons are adjacent to each other.

Problem 2.28 What is the area of the *smallest* equilateral triangle that can be inscribed in a 1-by-1 square? Note: Inscribed here means all 3 vertices of the triangle are on the square.

Problem 2.29 Let $ABCD$ be a unit square and let P, Q be on sides $\overline{AD}, \overline{AB}$ respectively such that $\triangle APQ$ has perimeter 2. Find the measure of $\angle PCQ$.

Problem 2.30 Give an example of a convex quadrilateral $ABCD$ (complete with the measure of all angles and all sides) that satisfies the constraints: $\angle A = 60°$, $\angle C = 30°$, and $AB = AD$. (Recall we have shown that $AC^2 = BC^2 + CD^2$.)

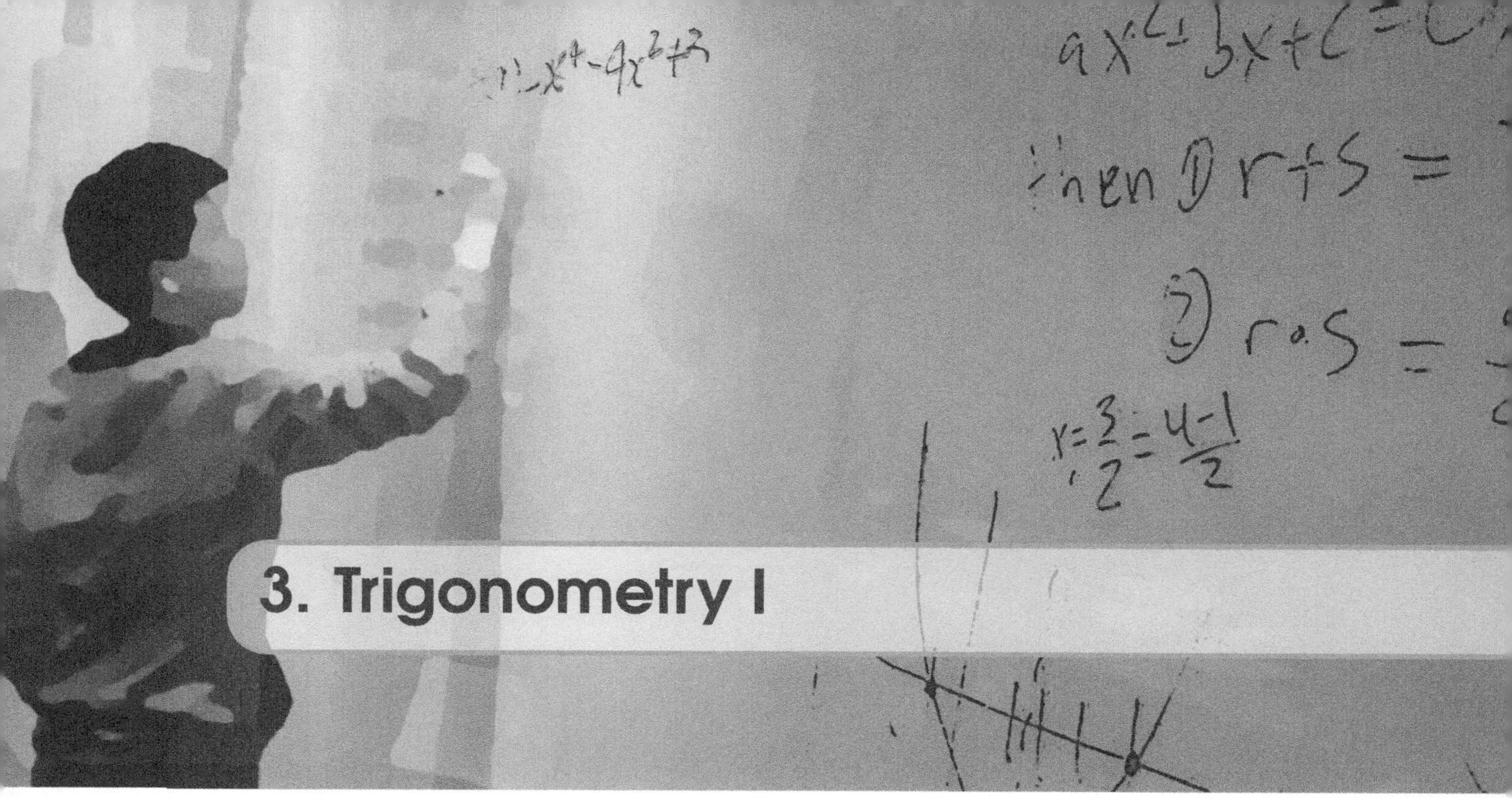

3. Trigonometry I

Trigonometry Review

- Definitions of Sine, Cosine, and Tangent
    - SOH-CAH-TOA: Sine is opposite over hypotenuse, Cosine is adjacent over hypotenuse, Tangent is opposite over adjacent
    - How can these be extended for all angles using the unit circle?
- Definitions of Secant, Cosecant, and Cotangent

The Special Angles

Radians	0	$\pi/6$	$\pi/4$	$\pi/3$	$\pi/2$	$2\pi/3$	$3\pi/4$	$5\pi/6$	π
Degrees	0	30	45	60	90	120	135	150	180
$\sin x$	0	1/2	$\sqrt{2}/2$	$\sqrt{3}/2$	1	$\sqrt{3}/2$	$\sqrt{2}/2$	1/2	0
$\cos x$	1	$\sqrt{3}/2$	$\sqrt{2}/2$	1/2	0	$-1/2$	$-\sqrt{2}/2$	$-\sqrt{3}/2$	-1
$\tan x$	0	$\sqrt{3}/3$	1	$\sqrt{3}$	-	$-\sqrt{3}$	-1	$-\sqrt{3}/3$	0
$\cot x$	-	$\sqrt{3}$	1	$\sqrt{3}/3$	0	$-\sqrt{3}/3$	-1	$-\sqrt{3}$	-
$\sec x$	1	$2\sqrt{3}/3$	$\sqrt{2}$	2	-	-2	$-\sqrt{2}$	$-2\sqrt{3}/3$	-1
$\csc x$	-	2	$\sqrt{2}$	$2\sqrt{3}/3$	1	$2\sqrt{3}/3$	$\sqrt{2}$	2	-

The Magic Hexagon (Wheel) of Trigonometry Identities

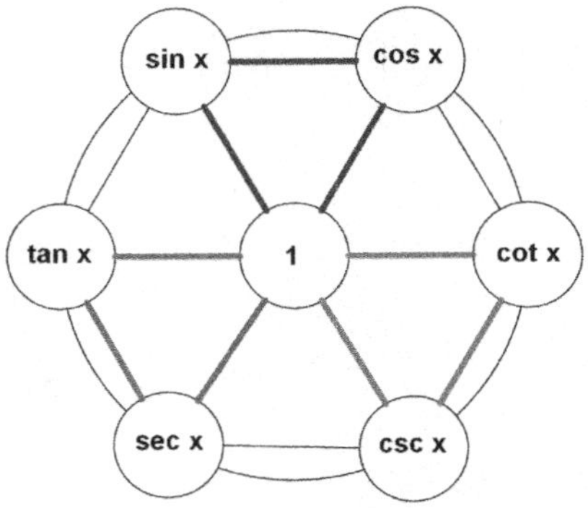

1. **On the diagonals—the reciprocal identities**

$$\sin x = \frac{1}{\csc x}, \qquad \cos x = \frac{1}{\sec x}, \qquad \tan x = \frac{1}{\cot x}$$

2. **On the upside-down triangles—the pythagorean identities**

$$\begin{aligned} \sin^2 x &+ \cos^2 x &= 1 \\ \tan^2 x &+ 1 &= \sec^2 x \\ 1 &+ \cot^2 x &= \csc^2 x \end{aligned}$$

3. **On the circle—the quotient identities** Any vertex is equal to the quotient of the next two consecutive vertices. It works in both directions.

$$\tan x = \frac{\sin x}{\cos x}, \qquad \sin x = \frac{\cos x}{\cot x}, \qquad \sec x = \frac{\tan x}{\sin x}, \qquad \text{etc.}$$

Theorems

- **Law of Cosines:** In $\triangle ABC$, let $BC = a, CA = b, AB = c$, and then

$$\begin{aligned} a^2 &= b^2 + c^2 - 2bc\cos A \\ b^2 &= c^2 + a^2 - 2ca\cos B \\ c^2 &= a^2 + b^2 - 2ab\cos C \end{aligned}$$

- **Law of Sines:** In $\triangle ABC$, let $BC = a, CA = b, AB = c$, and then $\frac{a}{\sin A} = \frac{b}{\sin B} = \frac{c}{\sin C}$.

3.1 Example Questions

Problem 3.1 In the diagram below, points E, B, C all lie on a unit circle with center O.

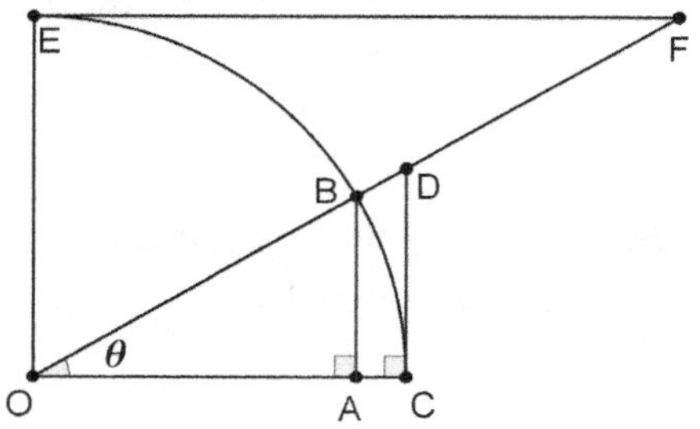

Calculate the following in terms of θ:

1. AO
2. AB
3. CD
4. DO
5. EF

Problem 3.2 Write the following in terms of $\sin(\theta), \cos(\theta)$, etc.

1. $\sin(90^\circ - \theta)$
2. $\cos(90^\circ - \theta)$
3. $\sin(180^\circ + \theta)$
4. $\cos(180^\circ + \theta)$
5. $\sin(180^\circ - \theta)$
6. $\cos(180^\circ - \theta)$
7. $\sin(-\theta)$
8. $\cos(-\theta)$

Problem 3.3 Prove the Law of Sines.

Problem 3.4 Let ABC be an equilateral triangle, and let P be a point in the interior of the triangle. Given that $PA = 3$, $PB = 4$, and $PC = 5$, find the side length of $\triangle ABC$.

Problem 3.5 In triangle ABC, $AB = c$, $BC = a$, and $CA = b$. Suppose that $(a+b+c)(a+b-c) = 3ab$. Determine the measure of $\angle C$.

Problem 3.6 A triangle has sides $\sqrt{3}-1 < x < \sqrt{6}$ and contains angles of 15° and 45°. Use this information to calculate $\sin(15^\circ)$.

Problem 3.7 Points A, B, C, D lie in the plane. Suppose C lies between A, B on $\overline{AB}$ while D is not on line $\overleftrightarrow{AB}$. If $CD = 4, BC = 4, AC = 5$, and $BD = 6$, what is AD?

Problem 3.8 Solve the following equations. Express your answers in radians.

(a) $2\tan(x) = \sec(x)$.

(b) $2\sin\left(2x - \frac{\pi}{2}\right) = \sqrt{2}$

(c) $\sin(x) + \sin(5x) - 2 = 0$.

Problem 3.9 Suppose $ABCD$ is a square with side length 2. Suppose a line containing A intersects side CD at E and the line extending BC at F. Find (with proof) the value of $\frac{1}{AE^2}+\frac{1}{AF^2}$.

Problem 3.10 Suppose you have a convex quadrilateral with diagonals c,d which intersect at an angle of θ. Find a formula for the area of the parallelogram.

3.2 Quick Response Questions

Problem 3.11 $\frac{5\pi}{8}$ radians is equal to how many degrees?

Problem 3.12 Which of the following is NOT equal to $\sin(45°)$?

(A) $\sqrt{2}/2$
(B) $-\cos(3\pi/4)$
(C) $\cos(-5\pi/4)$
(D) $\sin(-5\pi/4)$

Problem 3.13 Calculate $\cos(\pi/6) \cdot \tan(\pi/6)$. Round your answer to the nearest tenth if necessary.

Problem 3.14 Let θ be the smallest angle in a triangle with sides $5, 12, 13$. What is $\cot(\theta)$? Round your answer to the nearest hundredth if necessary.

Problem 3.15 $\triangle ABC$ has sides $BC = 5$ with $\angle A = 30°, \angle B = 45°$. What is side AC? Round your answer to the nearest tenth if necessary.

Problem 3.16 Which of the following is equal to $\dfrac{\cos(\theta)+1}{\cos(\theta)-1}$?

(A) $\dfrac{1-\sec(\theta)}{1-\sec(\theta)}$

(B) $\dfrac{1-\sec(\theta)}{1-\sec(\theta)}$

(C) $\dfrac{1+\sec(\theta)}{1+\sec(\theta)}$

(D) $\dfrac{1+\sec(\theta)}{1-\sec(\theta)}$

Problem 3.17 Write $\dfrac{\sec(\theta)}{1+\tan^2(\theta)}$ as a single trig. function.

(A) $\cos(\theta)$
(B) $\sin(\theta)$
(C) $\csc(\theta)$
(D) $\sec(\theta)$

Problem 3.18 Suppose a triangle has angles $15^\circ, 30^\circ, 135^\circ$. If the side opposite from the 30° angle is 2, then the length of the side opposite from the 135° angle can be written as $\sqrt{D}$ for an integer D. What is D?

Problem 3.19 Suppose you have a $5,6,7$ triangle with smallest angle θ. Then $\cos(\theta) = \dfrac{P}{Q}$ for integers P,Q with $Q>0$ and $\gcd(P,Q)=1$. What is $P+Q$?

Problem 3.20 $\triangle ABC$ has sides $BC = a, AC = b, AB = c$. Then $a^2 + b^2 + c^2$ is equal to which of the following?

(A) $bc\cos(A) + ac\cos(B) + ab\cos(C)$
(B) $2(bc\sin(A) + ac\sin(B) + ab\sin(C))$
(C) $2(bc\cos(A) + ac\cos(B) + ab\cos(C))$
(D) $a\sin(A) + b\cos(B) + c\cos(C)$

3.3 Practice Questions

Problem 3.21 Suppose $0^\circ < \theta < 90^\circ$ and $\cos(\theta) = \frac{7}{25}$. Calculate:

1. $\sin(\theta)$.
2. $\tan(\theta)$.
3. $\sec(\theta)$.
4. $\csc(\theta)$.
5. $\cot(\theta)$.

Problem 3.22 Write the following in terms of $\sin(\theta), \cos(\theta)$, etc.

1. $\tan(90^\circ - \theta)$
2. $\tan(180^\circ + \theta)$
3. $\tan(180^\circ - \theta)$
4. $\tan(-\theta)$
5. $\sin(360^\circ - \theta)$
6. $\cos(360^\circ - \theta)$
7. $\tan(360^\circ - \theta)$

Problem 3.23 Prove the Law of Cosines.

Problem 3.24 Let ABC be an isosceles right triangle with $\angle C = 90^\circ$. Let P be a point inside the triangle such that $AP = 3, BP = 5$, and $CP = 2\sqrt{2}$. What is $[ABC]$?

Problem 3.25 In triangle ABC, $AB = c$, $BC = a$, and $CA = b$. Suppose that $(a+b+c)(a+b-c) = ab$. Determine the value of $\sin C$.

Problem 3.26 Suppose there is an isosceles triangle containing a 120 degree angle. If the side opposite the 120 degree angle has length $4\sqrt{3}$, what are the lengths of the remaining two sides?

Problem 3.27 Points A, B, C, D lie in the plane. Suppose C lies between A, B on $\overline{AB}$ while D is not on line $\overleftrightarrow{AB}$. If $CD = BC = 3, AC = 6$, and $BD = 4$, what is AD?

Problem 3.28 Solve $\sqrt{3}\sin(2x) = 3\cos(2x)$.

Problem 3.29 Suppose $ABCD$ is a square with side length 2. Suppose a line intersects side CD at E and side BC at F so that $\angle AEF = 90°$. Set $\angle DAE = \theta$. Write a formula for the area of $\triangle AEF$ in terms of θ.

Problem 3.30 Suppose you have a parallelogram with side lengths a, b and diagonals c, d. Prove that $2a^2 + 2b^2 = c^2 + d^2$.

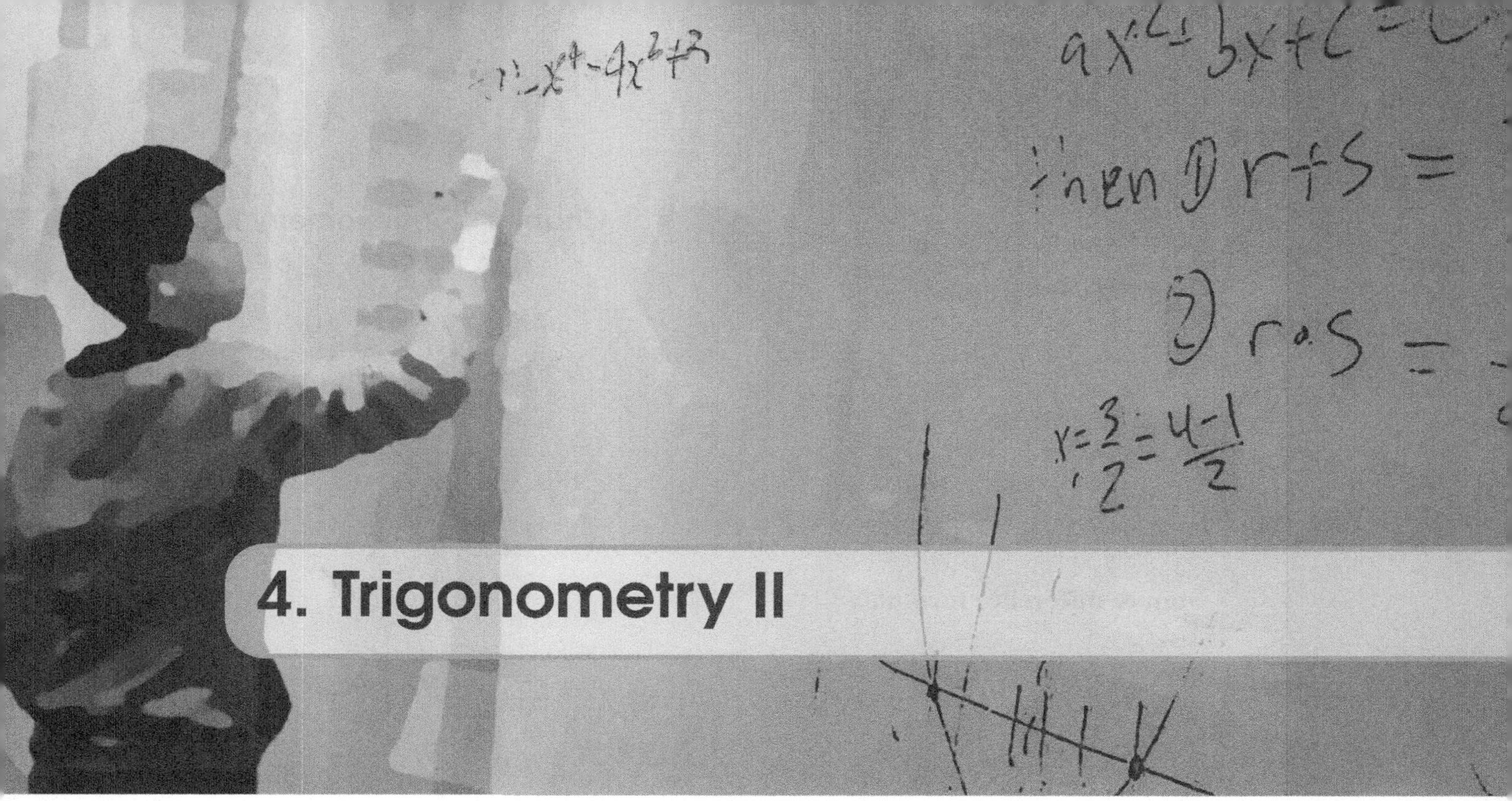

4. Trigonometry II

Review

- **Law of Cosines:** In $\triangle ABC$, let $BC = a, CA = b, AB = c$, and then

$$\begin{aligned} a^2 &= b^2 + c^2 - 2bc\cos A \\ b^2 &= c^2 + a^2 - 2ca\cos B \\ c^2 &= a^2 + b^2 - 2ab\cos C \end{aligned}$$

- **Law of Sines:** In $\triangle ABC$, let $BC = a, CA = b, AB = c$, and then $\dfrac{a}{\sin A} = \dfrac{b}{\sin B} = \dfrac{c}{\sin C}$.

Trigonometric Identites (some you've already proven, others will be proven in exercises or homework)

- **Co-functions (complementary angles)**

$$\sin(90^\circ - x) = \cos x, \qquad \cos(90^\circ - x) = \sin x, \qquad \tan(90^\circ - x) = \cot x$$

- **Even-Odd properties**

$$\sin(-x) = -\sin x, \qquad \cos(-x) = \cos x, \qquad \tan(-x) = -\tan x$$

- **Periodic properties**

$$\sin(x + 360^\circ) = \sin x, \qquad \cos(x + 360^\circ) = \cos x, \qquad \tan(x + 180^\circ) = \tan x$$

- **Supplementary angles**

$$\sin(180^\circ - x) = \sin x, \qquad \cos(180^\circ - x) = -\cos x, \qquad \tan(180^\circ - x) = -\tan x$$

- **Other properties**

$$\sin(x + 180^\circ) = -\sin x, \qquad \cos(x + 180^\circ) = -\cos x$$

$$\sin(360^\circ - x) = -\sin x, \qquad \cos(360^\circ - x) = \cos x, \qquad \tan(360^\circ - x) = -\tan x$$

- **Sum & difference formulas**

$$\begin{aligned}
\sin(x \pm y) &= \sin x \cos y \pm \cos x \sin y \\
\cos(x \pm y) &= \cos x \cos y \mp \sin x \sin y \\
\tan(x \pm y) &= \frac{\tan x \pm \tan y}{1 \mp \tan x \tan y}
\end{aligned}$$

- **Double angle formulas**

$$\begin{aligned}
\sin 2x &= 2 \sin x \cos x \\
\cos 2x &= \cos^2 x - \sin^2 x \\
&= 2\cos^2 x - 1 \\
&= 1 - 2\sin^2 x \\
\tan 2x &= \frac{2\tan x}{1 - \tan^2 x}
\end{aligned}$$

- **Power-reducing/Half angle formulas**

$$\begin{aligned}
\sin^2 x &= \frac{1 - \cos 2x}{2}, & \cos^2 x &= \frac{1 + \cos 2x}{2} \\
\tan^2 x &= \frac{1 - \cos 2x}{1 + \cos 2x}, & \tan x &= \frac{\sin 2x}{1 + \cos 2x} = \frac{1 - \cos 2x}{\sin 2x}.
\end{aligned}$$

4.1 Example Questions

Problem 4.1 Prove that $\sin(x+y) = \sin(x)\cos(y) + \cos(x)\sin(y)$. The diagram below may be helpful:

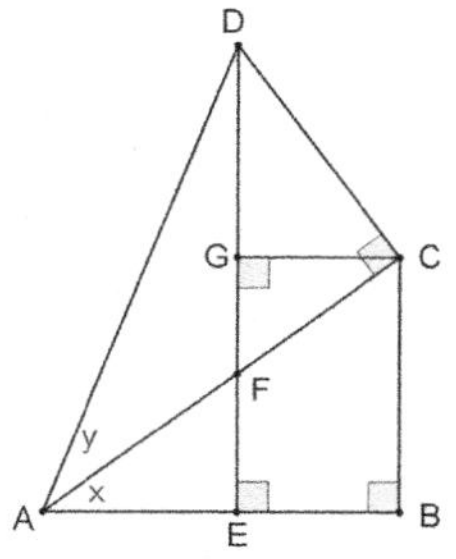

Hint: $DE = DG + GE$.

Problem 4.2 Prove the following. (Hint: Use previous formulas!)

(a) The sine difference formula.

(b) The sine double angle formula.

(c) The cosine double angle formulas.

Problem 4.3 Prove the following.

(a) The tangent sum and difference formulas.

(b) The tangent double angle formulas.

Problem 4.4 Compute the following values.

(a) $\cos(225^\circ)$.

(b) $\sin(75^\circ)$.

(c) $\cos(165^\circ)$.

(d) $\sec(525^\circ)$.

Problem 4.5 Simplify $\cos(A+B)\cos(B)+\sin(A+B)\sin(B)$.

Problem 4.6 In $\triangle ABC$, suppose $AB = 15$, $CD = 13$ and $\cos(B) = \frac{33}{65}$. What is $\sin(C)$?

Problem 4.7 Trig. Identities Practice

(a) Show that $\sin(2x)+\sin(2y) = 2\sin(x+y)\cos(x-y)$.

(b) Given $\triangle ABC$ (sides a,b,c), prove $b\cos(B)+c\cos(C) = a\cos(B-C)$.

Problem 4.8 Find all solutions x (in radians) so that $0 \leq x < 2\pi$ to the following equations.

(a) $\sin(2x) = \tan(x)$.

(b) $\cos(2x) = \cos(x)$.

(c) $\sin(x) + \cos(x) = \frac{\sqrt{6}}{2}$. Hint: $\cos(\pi/4) = \sin(\pi/4) = \sqrt{2}/2$.

Problem 4.9 Suppose in triangle ABC we have $\angle A = 45^\circ$, $AB = 3$ and $AC = 2\sqrt{2}$. What is $\tan(B)$?

Problem 4.10 Show that in a triangle ABC (with sides a, b, c) we have $\dfrac{a-b}{a+b} = \tan\left(\dfrac{A-B}{2}\right)\tan\left(\dfrac{C}{2}\right)$. Hints: Start by writing the left hand side in terms of sines. An identity we proved earlier might be useful again here!

4.2 Quick Response Questions

Problem 4.11 Find the exact value of $\sin(15^\circ)$.

(A) $\frac{\sqrt{6}-\sqrt{2}}{2}$
(B) $\frac{\sqrt{6}-\sqrt{2}}{4}$
(C) $\frac{\sqrt{2}-1}{2}$
(D) $\frac{\sqrt{6}+\sqrt{2}}{4}$

Problem 4.12 Find the exact value of $\tan(15^\circ)$.

(A) $\frac{3-\sqrt{3}}{3}$
(B) $\frac{3+\sqrt{3}}{3}$
(C) $2+\sqrt{3}$
(D) $2-\sqrt{3}$

Problem 4.13 Which of the following is not equal to $\sin(120^\circ)$?

(A) $2\sin(60^\circ)\cos(60^\circ)$
(B) $1-2\sin^2(15^\circ)$
(C) $2\cos^2(60^\circ)-1$
(D) $\sin(30^\circ)\cos(90^\circ)+\cos(30^\circ)\sin(90^\circ)$

Problem 4.14 Which of the following is not equal to $\tan(60°)$?

(A) $\dfrac{\sin(120°)}{1-\cos(120°)}$
(B) $\dfrac{2\tan(30°)}{1-\tan^2(30°)}$
(C) $-\tan(120°)$
(D) $\dfrac{1-\cos(120°)}{\sin(120°)}$

Problem 4.15 Simplify $\dfrac{\cos(t+180°)\sin(t+360°)}{\sin(-t-180°)\cos(-t-180°)}$.

(A) $\cos(t)$
(B) 1
(C) -1
(D) $-\sin(t)$

Problem 4.16 Which of the following is equal to $1+\sin(2\theta)$?

(A) $(\cos(\theta)+\sin(\theta))^2$
(B) $\cos(\theta)+\sin(\theta)$
(C) $\cos^2(\theta)+\cos^2(\theta)$
(D) $\cos^2(\theta)-\sin^2(\theta)$

Problem 4.17 Which of the following is equal to $\sec^2(\theta) - 1$?

(A) $\tan(\theta)$

(B) $\dfrac{1-\cos(2\theta)}{1+\cos(2\theta)}$

(C) $\dfrac{\sin(2\theta)}{1+\cos(2\theta)}$

(D) $\dfrac{1+\cos(2\theta)2}{2}$

Problem 4.18 There is one solution (in degrees) with $0^\circ < \theta < 360^\circ$ to $\cos(\theta) + \sin(\theta) = 1$. What is this solution?

Problem 4.19 Suppose $\sin(\theta) = \dfrac{5}{13}$ where $0 < \theta < \dfrac{\pi}{2}$. Then $\sin(2\theta) = \dfrac{P}{Q}$ with $Q > 1$ and $\gcd(P, Q) = 1$. What is $P + Q$?

Problem 4.20 Suppose $\sin(\theta) = \dfrac{5}{13}$ where $\dfrac{\pi}{2} < \theta < \pi$. Then $\tan(2\theta) = \dfrac{P}{Q}$ with $Q > 1$ and $\gcd(P, Q) = 1$. What is $P + Q$?

4.3 Practice Questions

Problem 4.21 Prove that $\cos(x+y) = \cos(x)\cos(y) - \sin(x)\sin(y)$. Hint: Recall the diagram from the sine sum formula proved in class.

Problem 4.22 Prove the following. (Hint: Use previous formulas!)

(a) The cosine difference formula.

(b) The sine power-reducing formula.

(c) The cosine power-reducing formula.

Problem 4.23 Prove the tangent power-reducing formula.

Problem 4.24 Compute the following values.

(a) $\sin(330^\circ)$.

(b) $\tan(405^\circ)$.

(c) $\tan(75^\circ)$.

Problem 4.25 Simplify $\sin(x)\cos^3(x) - \cos(x)\sin^3(x)$.

Problem 4.26 Given $\triangle ABC$ with $\sin(B) = \frac{24}{25}, \sin(C) = \frac{12}{13}$, calculate $\sin(A)$.

Problem 4.27 In $\triangle ABC$ (with sides a, b, c), show that $b\cos(C) + c\cos(B) = a$.

Problem 4.28 Find all solutions x (in radians) so that $0 \le x < 2\pi$ to the following equations.

(a) $\sin(2x) = \sqrt{3}\sin(x)$.

(b) $\cos(2x) = \sin(x)$.

Problem 4.29 Suppose in triangle ABC we have $\angle A = 45^\circ$, $AB = 3$ and $AC = 2\sqrt{2}$. Calculate $\tan(C)$.

Problem 4.30 In triangle ABC, suppose $\angle A = 60^\circ$ and $\frac{b}{c} = 2 + \sqrt{3}$. Calculate $\angle B$ and $\angle C$. Hint: Recall Problem 4.10 and use it to calculate $\tan((B - C)/2)$.

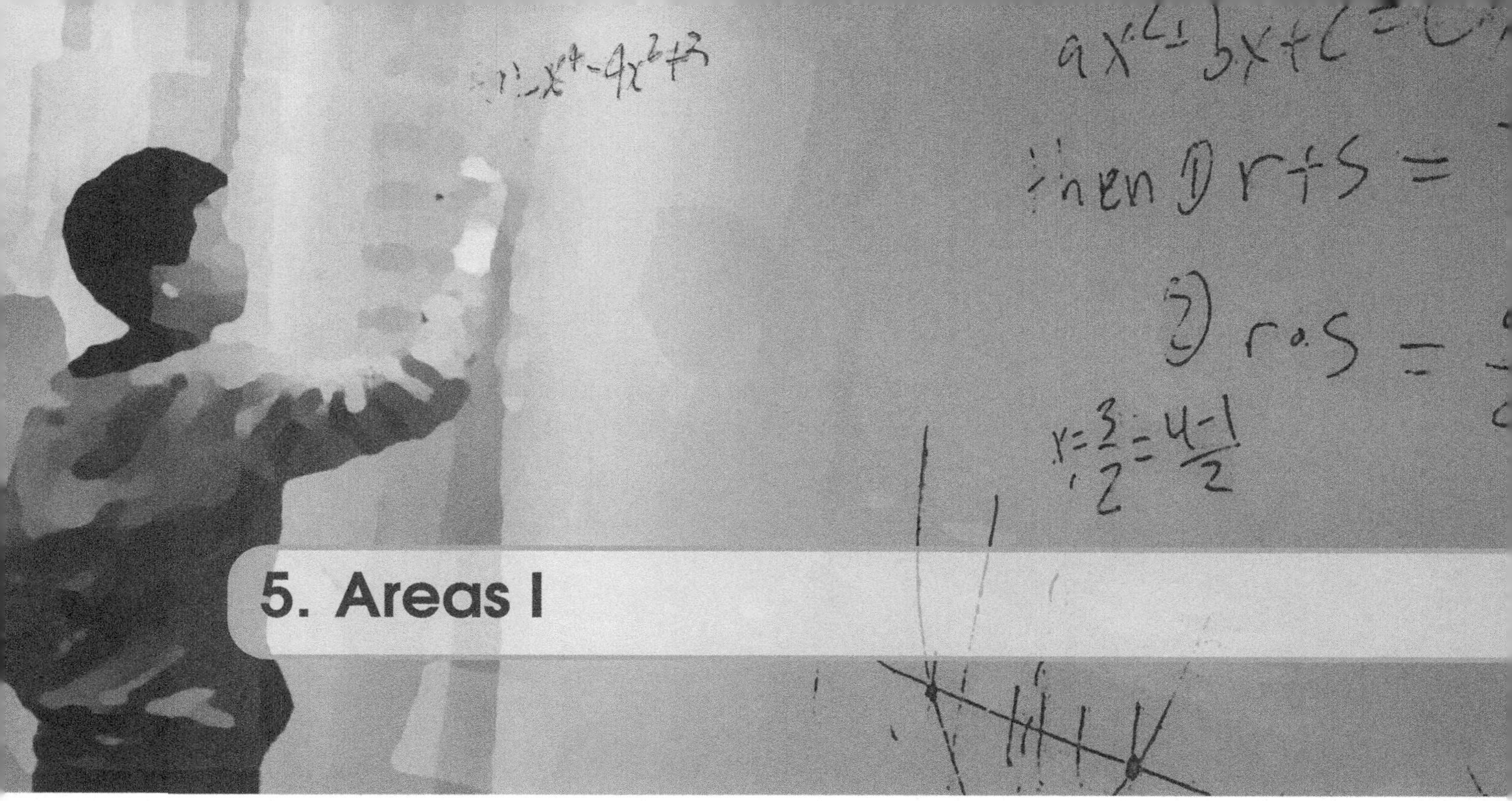

5. Areas I

Concepts and Facts

- **Notation:** We shall use $[ABC]$ to denote the area of triangle ABC, $[XYZW]$ to denote the area of the quadrilateral $XYZW$, etc.
- Formulas for areas (should be memorized): square, rectangle, triangle, parallelogram, trapezoid, circle

  Various area formulas of triangle ABC:

$$
\begin{aligned}
[ABC] &= \frac{1}{2}ah && (h \text{ is the altitude on } a) \\
&= \frac{abc}{4R} && (R \text{ is the circumradius}) \\
&= \sqrt{s(s-a)(s-b)(s-c)} && (s = \frac{a+b+c}{2}) \\
&= rs && (r \text{ is the inradius, } s \text{ is defined as above})
\end{aligned}
$$

- The areas of triangles (or parallelograms) with equal bases and equal altitudes (heights) are equal.
- The areas of triangles with equal altitudes are proportional to the bases of the triangles.
- The ratio of areas between two similar triangles is the square of the ratio between the corresponding sides.

5.1 Example Questions

Problem 5.1 Using only the basics about parallel lines and congruent/similar triangles and the fact that the area of a rectangle is bh, prove the following. (Note: once a fact is proven below, you can use it in later parts.)

(a) The area of a parallelogram is bh.

(b) The area of a triangle is $\frac{1}{2}bh$ (prove this two ways!).

(c) The area of a trapezoid is $\frac{b_1+b_2}{2}h$.

(d) The area of a trapezoid is also mh where m is the *median* of the trapezoid, which connects the midpoints of the two non-parallel sides.

Problem 5.2 What is the ratio of areas between two similar triangles? Prove your result!

Problem 5.3 Prove the Pythagorean Theorem using areas.

Problem 5.4 In $\triangle ABC$, $AB > AC > BC$, $\overline{CD}, \overline{BE}, \overline{AF}$ are altitudes on $\overline{AB}, \overline{AC}, \overline{BC}$, respectively. Show that $CD < BE < AF$.

Problem 5.5 In triangle ABC, $AC = 10$, $BC = 24$, $AB = 26$. What is the altitude on $\overline{AB}$?

Problem 5.6 Let $ABCD$ be a parallelogram, and E,H,F,G be points on sides $\overline{AB},\overline{BC},\overline{CD},\overline{DA}$ respectively, and $\overline{EF}||\overline{BC}$ and $\overline{GH}||\overline{AB}$. Let P be the intersection of $\overline{EF}$ and $\overline{GH}$. If $[GPFD] = 10, [PHCF] = 8, [EBHP] = 16$, find $[ABCD]$.

Problem 5.7 Suppose you only know that the centroid exists. Prove (using areas!) that the centroid divides each median in a ratio of 1 : 2.

Problem 5.8 Let $ABCD$ be a parallelogram, with midpoints E,F,G,H (say on $\overline{AB},\overline{BC},\overline{CD},\overline{DA}$). Let I,J be the midpoints of $\overline{EF},\overline{GH}$. Find the area of $\triangle JIG$ as a fraction of the area of $ABCD$.

Problem 5.9 Prove that if a triangle has side lengths a,b,c, inradius r, and circumradius R we have $2Rr = \dfrac{abc}{a+b+c}$.

Problem 5.10 Let $ABCD$ be a parallelogram as in the diagram, with E the midpoint of $\overline{BC}$.

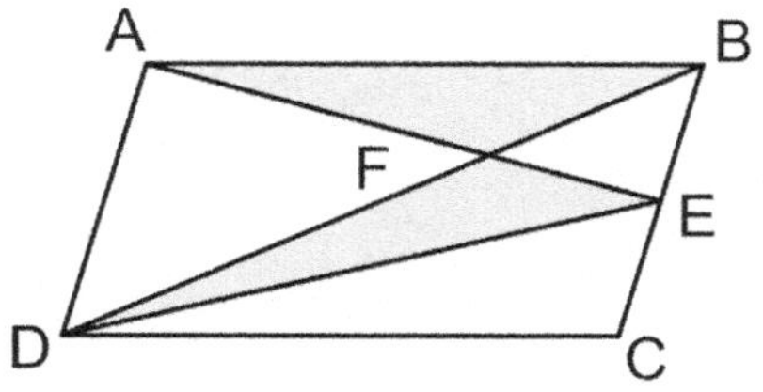

(a) Compare the shaded regions $\triangle ABF$ and $\triangle DEF$, which one has the larger area?

(b) Find the area of the shaded regions $\triangle ABF$ and $\triangle DEF$ in terms of the entire area of the parallelogram.

5.2 Quick Response Questions

Problem 5.11 If $AB \parallel CD$, can we conclude that $[ABC] = [ABD]$?

Problem 5.12 Consider the two squares in the diagram below.

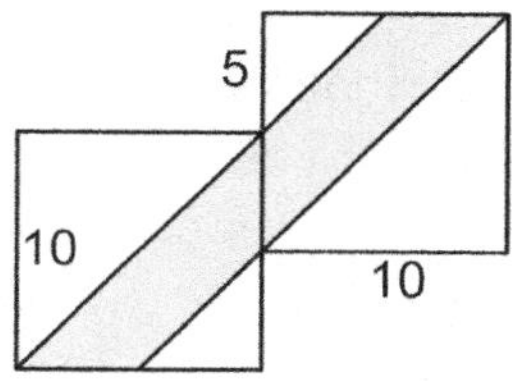

Find the area of the shaded region.

Problem 5.13 Consider two squares attached at their corner vertices as in the diagram below.

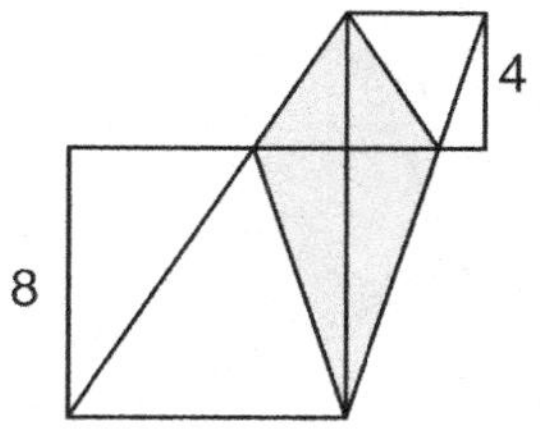

Find the area of the shaded region.

Problem 5.14 Suppose $\triangle ABC$ with E on $\overline{AB}$ and D on $\overline{AC}$ such that $AE = AB/3, AD = AC/2$. If $[AED] = 2$, find the area of $[ABC]$.

Problem 5.15 In triangle ABC, $AC = 9$, $BC = 10$, $AB = 17$. What is the altitude on $\overline{BC}$?

Problem 5.16 $\triangle ABC$ has area 54. Let D be a point on AB such that $AD = 2DB$. What is $[ADC]$?

Problem 5.17 Triangle ABC is inscribed on a circle of radius 10. $AB = 19$, $AC = 12$, and $[ABC] = 65.25$. What is BC? Round your answer to the nearest hundredth.

Problem 5.18 Let $\triangle ABC$ be an isosceles triangle with $AC = BC$. Let D, E and F be the midpoints of BC, CA and AB, respectively and let G be its centroid. If $EB = 12$, what is AG?

Problem 5.19 Consider the parallelogram $ABCD$. Let E and F be midpoints of the sides CD and DA, respectively. If $[ABCD] = 40$, what is $[BEF]$?

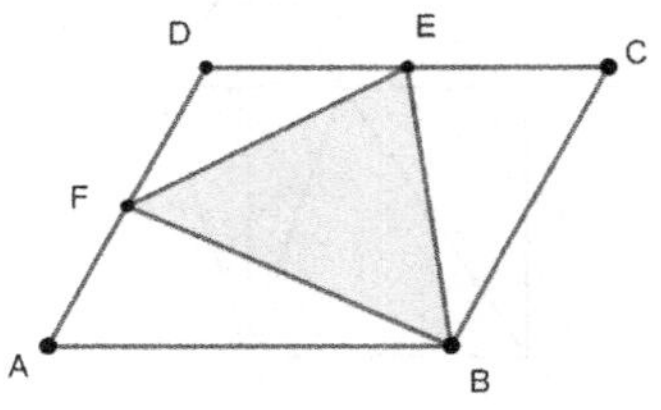

Problem 5.20 In the following diagram, G is the centroid of $\triangle ABC$. If $[ABC] = 92$, what is the shaded area?

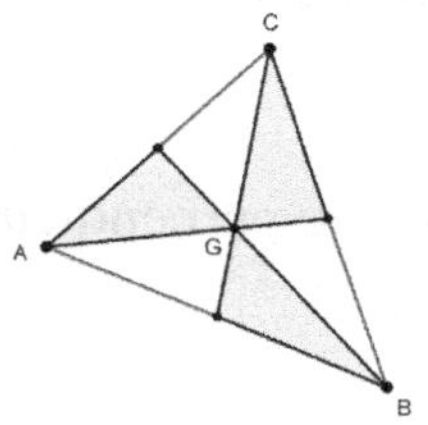

5.3 Practice Questions

Problem 5.21 Recall that a kite is a quadrilateral with two sets of adjacent sides of equal length.

(a) Prove that the diagonals in a kite are perpendicular.

(b) Prove that the area of a kite is $\frac{d_1 \cdot d_2}{2}$ (where d_1, d_2 are the diagonals) using the same ideas as Problem 5.1.

Problem 5.22 In $\triangle ABC$, let D, E, F be midpoints of the sides $\overline{BC}, \overline{AC}, \overline{AB}$. Show that $[DEF] = [ABC]/4$.

Problem 5.23 Prove the converse of the Pythagorean Theorem. Hint: You can use the Pythagorean Theorem!

Problem 5.24 Let $ABCD$ be a parallelogram with $[ABCD] = 1$. Let P be a point in the interior of $ABCD$. Show that $[ABP] + [CDP]$ is a fixed value and find that value.

Problem 5.25 Find a formula for the area of a parallelogram if you are given the two sides lengths as well as one of the diagonals.

Problem 5.26 Let $ABCD$ be a rectangle, and E,H,F,G be points on sides $\overline{AB},\overline{BC},\overline{CD},\overline{DA}$ respectively, and $\overline{EF}||\overline{BC}$ and $\overline{GH}||\overline{AB}$. Let P be the intersection of EF and GH. If $[GPFD]=10,[EBHP]=12$ and $[ABCD]=44$. Suppose each of the four rectangles formed above has integer dimensions, find all possible dimensions for $GPFD$ and $EBHP$.

Problem 5.27 Prove that the centroid is well-defined using areas (that is, show the medians are concurrent). Hint: Your proof will be an extension of the one done in class.

Problem 5.28 In parallelogram $ABCD$, M and N are midpoints of sides $\overline{AB}$ and $\overline{BC}$ respectively. Given that $[DMN]=9$, find the area of $ABCD$.

Problem 5.29 Prove that if a triangle has side lengths a,b,c semiperimeter s, inradius r, and circumradius R we have

$$\frac{R}{r}=\frac{abc}{4(s-a)(s-b)(s-c)}.$$

Problem 5.30 Suppose a parallelogram P_1 has area 256. Connect the midpoints of each side to form a parallelogram P_2. Repeat to get P_3, and continue repeating until you get to P_{10}.

(a) Prove this problem actually makes sense. That is, prove that if you connect the midpoints of a parallelogram you get another parallelogram.

(b) Find the area of P_{10}.

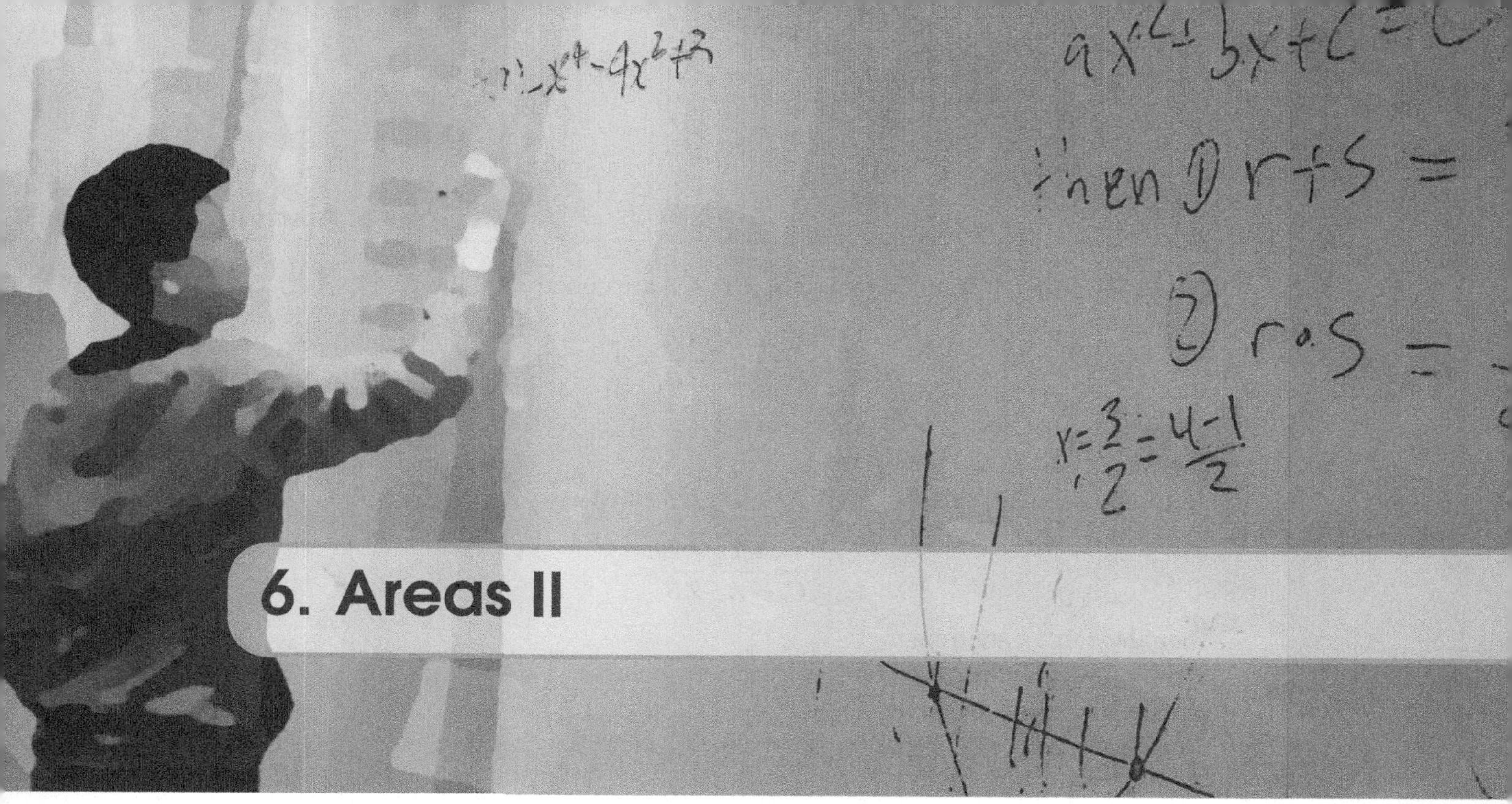

6. Areas II

Concepts and Facts

- **Notation:** We shall use $[ABC]$ to denote the area of triangle ABC, $[XYZW]$ to denote the area of the quadrilateral $XYZW$, etc.
- Formulas for areas (should be memorized): square, rectangle, triangle, parallelogram, trapezoid, circle
  Various area formulas of triangle ABC:

$$\begin{aligned}
[ABC] &= \frac{1}{2}ah && (h \text{ is the altitude on } a) \\
&= \frac{abc}{4R} && (R \text{ is the circumradius}) \\
&= \sqrt{s(s-a)(s-b)(s-c)} && (s = \frac{a+b+c}{2}) \\
&= rs && (r \text{ is the inradius, } s \text{ is defined as above})
\end{aligned}$$

- The areas of triangles (or parallelograms) with equal bases and equal altitudes (heights) are equal.
- The areas of triangles with equal altitudes are proportional to the bases of the triangles.
- The ratio of areas between two similar triangles is the square of the ratio between the corresponding sides.

Theorems

- **Ceva's Theorem** ("Ceva" is pronounced "chayva")

In triangle ABC, if three cevians AX, BY, CZ are concurrent, then

$$\frac{BX}{XC} \cdot \frac{CY}{YA} \cdot \frac{AZ}{ZB} = 1.$$

Proof: Use areas (see questions below).

- **Converse of Ceva's Theorem**
  In triangle ABC, if three cevians AX, BY, CZ satisfy

  $$\frac{BX}{XC} \cdot \frac{CY}{YA} \cdot \frac{AZ}{ZB} = 1,$$

  then they are concurrent.
  Proof: Use an indirect proof (see questions below).
- **Angle Bisector Theorem**
  In triangle ABC, let D be the point on $\overline{BC}$ such that $\overline{AD}$ bisects $\angle BAC$. Then $AB/AC = BD/DC$.
  Proof: There are many different proofs using similar triangles. Try using an area argument (see questions below).

6.1 Example Questions

Problem 6.1 Ceva's Theorem and Converse.

(a) Ceva's Theorem and Its Converse

(b) Prove the converse to Ceva's Theorem using an indirect proof.

Problem 6.2 (Centers of Triangles) Prove the following, using Ceva's Theorem (or its converse) if applicable.

(a) (Centroid) In every triangle ABC, the three medians (i.e. the line from a vertex to the midpoint of the opposite side) are concurrent. This point is called the *centroid* of the triangle ABC.

(b) (Incenter) In every triangle ABC, the three angle bisectors (i.e. the line from a vertex bisecting the angle at that vertex) are concurrent. This point is called the *incenter* of the triangle ABC, because it is the center of the circle inscribed in triangle ABC.

Problem 6.3 Suppose you have a trapezoid $ABCD$ with $\overline{AB}$ parallel to $\overline{CD}$. Let E be the intersection of the diagonals. Suppose $AB = 10, CD = 15$ and $\triangle ADE$ has area 24. Find the area of $ABCD$.

Problem 6.4 Suppose the altitudes of a triangle are in ratio $2:2:3$ and the triangle has a perimeter of 24. Find the area of the triangle.

Problem 6.5 Let G be the centroid of $\triangle ABC$, and $AG = 3, BG = 4, CG = 5$, find $[ABC]$.

Problem 6.6 (2002 AMC 12A #22) Triangle ABC is a right triangle with $\angle ACB$ as its right angle, $m\angle ABC = 60^\circ$, and $AB = 10$. Let P be randomly chosen inside $\triangle ABC$, and extend $\overline{BP}$ to meet $\overline{AC}$ at D. What is the probability that $BD > 5\sqrt{2}$?

Problem 6.7 Suppose a dodecagon is in a square as in the diagram below (the vertices of the dodecagon on the square are the midpoints of the sides):

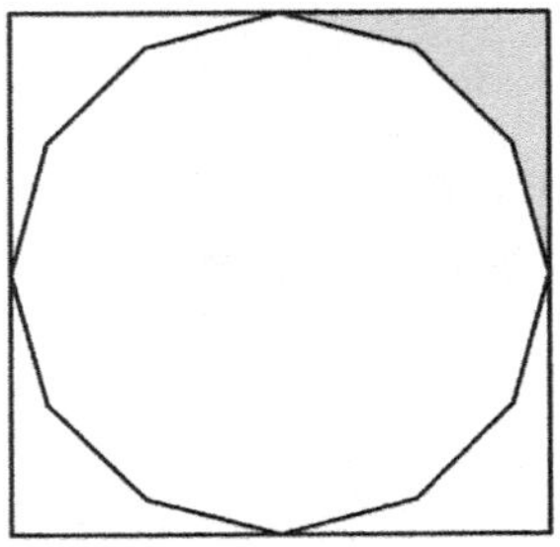

If the area of the square is 4, what is the area of the shaded region?

Problem 6.8 Let $ABCDE$ be a convex pentagon. Suppose further that the triangle cut off by each diagonal has area 1. What is the area of the full pentagon $ABCDE$?

Problem 6.9 Start with trapezoid $ABCD$. Extend $\overline{AD}, \overline{BC}$ to meet at O as in the diagram below.

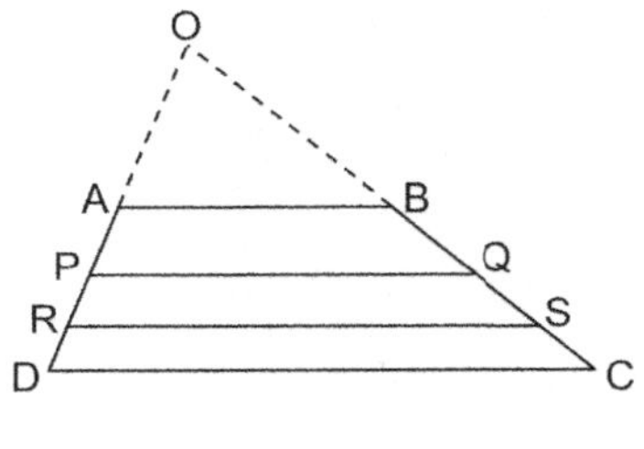

Suppose $AD = 1, OA = 1$. Construct P, Q, R, S such that $\overline{AB} \parallel \overline{PQ} \parallel \overline{RS}$ and $[ABQP] = [QSRP] = [SCDR]$. What are AP, PR, RD?

Problem 6.10 (Kurrah's Theorem) Let $\triangle ABC$ be given. Construct D, E as in the diagram below with $\angle ABC = \angle ADB = \angle BEC$.

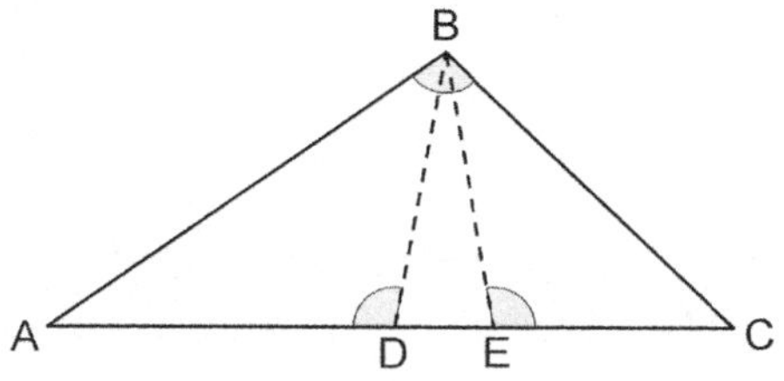

Prove that

$$AB^2 + BC^2 = AC(AD + CE).$$

6.2 Quick Response Questions

Problem 6.11 Suppose $\triangle ABC$ with E on $\overline{AB}$ and D on $\overline{AC}$ such that $AE = AB/3, AD = AC/2$. If $[AED] = 2$, find the area of $[ABC]$.

Problem 6.12 In the diagram below, there are 21 grid points arranged in equilateral triangles, equally spaced. The *area* of each small equilateral triangle formed by 3 adjacent grid points is 1. Find the area of $\triangle ABC$.

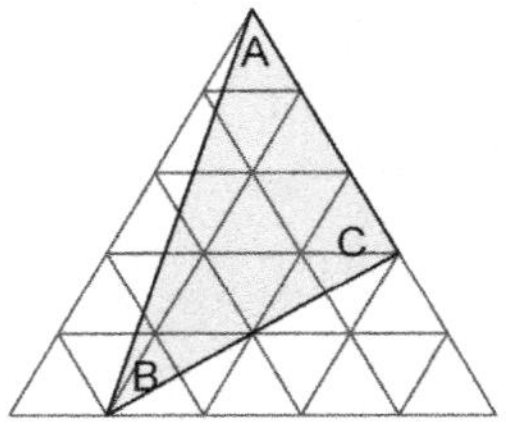

Problem 6.13 Suppose $ABCD$ is a parallelogram and P is an arbitrary point on the interior of $ABCD$. Which of the following is true?

(A) $[ABP] = [ABCD]/4$
(B) $[ABP] = [CDP]$
(C) $[ABP] + [CDP] = [ABCD]/2$
(D) None of the above

Problem 6.14 Let P be an interior point in parallelogram $ABCD$, and $[APB] : [ABCD] = 2 : 5$. $[CPD] : [ABCD] = a : b$, where a and b have no common factors. What is $a + b$?

Problem 6.15 Suppose you have a circle with diameter $\overline{AB}$ with $AB = 4$. Let C, D be on arc $\widehat{AB}$ such that $\widehat{AC} : \widehat{CD} : \widehat{DB} = 1 : 2 : 1$. Find the area of the figure enclosed by line segment $\overline{AC}$, arc $\widehat{CD}$, and line segment $\overline{AD}$. Use $\pi = 3.14$ and round your answer to the nearest tenth if necessary.

Problem 6.16 In $\triangle ABC$, $AB = 7$, $BC = 10$ and $AC = 14$. If $\angle ABC = \angle ADB = \angle CEB$, $DE = \frac{P}{Q}$ in lowest terms. What is $P + Q$?

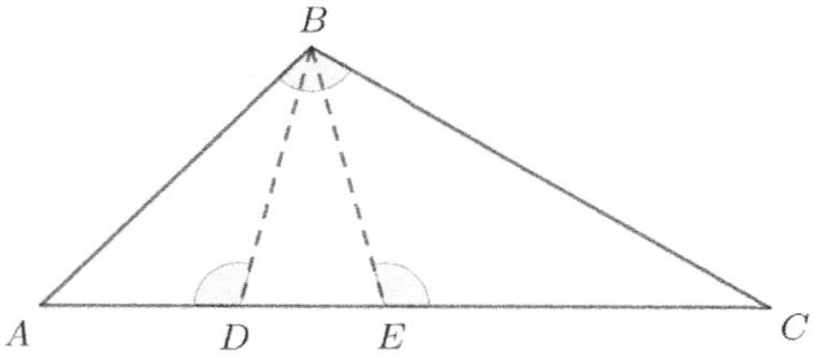

Problem 6.17 Suppose a dodecagon is in a square as in the diagram below (the vertices of the dodecagon on the square are the midpoints of the sides):

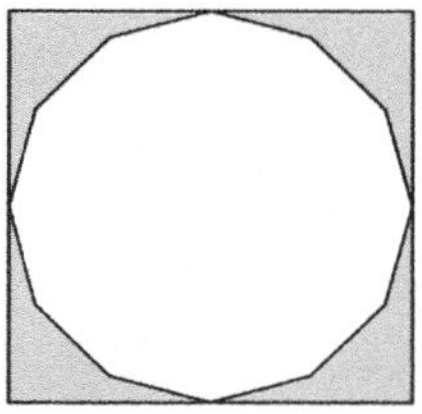

If the area of the square is 8, what is the area of the shaded region?

Problem 6.18 A triangle has side lengths 6, 8 and 12. The area of the triangle is $\sqrt{A}$, what A?

Problem 6.19 Consider the parallelogram $ABCD$. Let E be the middle point of AB and F be the middle point of CD. If $[ABCD] = 248$, what is $[EBF]$?

Problem 6.20 Let M be the intersection of the line segments $\overline{AB}$ and $\overline{PQ}$. If $PM : QM = 2 : 5$ and $[PAB] = 16$, what is $[QAB]$?

6.3 Practice Questions

Problem 6.21 Prove the Angle Bisector Theorem and its converse. That is, assume ABC is a triangle with D on $\overline{BC}$. Prove that $\overline{AD}$ bisects $\angle BAC$ if and only if $\dfrac{AB}{AC} = \dfrac{BD}{DC}$.

Problem 6.22 (Centers of Triangles Continued) Prove the following, using Ceva's Theorem (or its converse) if applicable.

(a) (Orthocenter) In every triangle ABC, the three altitudes (i.e. the line from a vertex perpendicular to the opposite side) are concurrent. The common point of intersection is called the *orthocenter* of triangle ABC.

(b) (Circumcenter). In every triangle ABC, the perpendicular bisectors of the three sides are concurrent. This point is called the *circumcenter* of the triangle ABC, because it is the center of the circle circumscribed around triangle ABC.

Problem 6.23 Given square $ABCD$, let E,F be the midpoints of AB and BC respectively, and G be the intersection of AF and CE. If $[ABCD] = 1$, find $[AGCD]$.

Problem 6.24 Suppose that a triangle has altitudes with ratio $12 : 15 : 20$ and area 192. Find the perimeter of the triangle.

Problem 6.25 ABC is a triangle with integer side lengths. Extend $\overline{AC}$ beyond C to point D such that $CD = 120$. Similarly, extend $\overline{CB}$ beyond B to point E such that $BE = 112$ and $\overline{BA}$ beyond A to point F such that $AF = 104$. If triangles CBD, BAE, and ACF all have the same area, what is the minimum possible area of triangle ABC?

Problem 6.26 (2002 AMC 12A #23) In triangle ABC, side $\overline{AC}$ and the perpendicular bisector of $\overline{BC}$ meet in point D, and $\overline{BD}$ bisects $\angle ABC$. If $AD = 9$ and $DC = 7$, what is the area of triangle ABD?

Problem 6.27 What is the area of a dodecagon that is inscribed in a circle of radius 1?

Problem 6.28 Suppose $ABCD$ is a trapezoid with height 8. If the diagonals are perpendicular and one diagonal has length 10, what is the area of the trapezoid?

Problem 6.29 Let $\triangle ABC$. Construct $\overline{DE} \parallel \overline{FG} \parallel \overline{BC}$ (with D on $\overline{AB}$ and F on $\overline{DB}$) dividing the triangle into three regions of the same area. What is $AD : DF : FB$?

Problem 6.30 Explain why the Pythagorean theorem follows almost immediately from Kurrah's theorem.

7. Circles I

Basic Definitions

- A *circle* is a collection of points of equal distance (called the *radius*) from a set point (called the *center*).
- Given two points A, B on a circle, the segment $\overline{AB}$ is called a *chord*.
- If a chord AB contains the center of the circle, we say A and B are *diametrically opposite*, and call AB a *diameter*.
- The portion of a circle that lies above or below a chord AB is called an *arc*. If the arc is more than half a circle it is called a *major arc*, less than half a circle is called a *minor arc*, and half a circle is called a *semicircle*. The arc will be denoted $\widehat{AB}$.
- Suppose $\widehat{AB}$ is an arc on a circle with center O. The *angular size* of the arc $\widehat{AB}$ is equal to the angle $\angle AOB$ (which is referred to as a *central angle*).
- Given a central angle $\angle AOB$ from an arc $\widehat{AB}$, the figure contained between the arc $\widehat{AB}$ and the radii $\overline{OA}, \overline{OB}$ is called a *sector*.

Measurements in Circles

- The area of a circle is given by πr^2 where r is the radius.
- The circumference of a circle is given by $2\pi r = \pi d$ where r, d are the radius and length of a diameter respectively.
- The *arc length* of $\widehat{AB}$ (that is, the distance walking from A to B along the circle) is given by $\frac{\theta}{360^\circ} 2\pi r$ where θ is the angular size of $\widehat{AB}$ (measured in degrees).

- Similarly, the area of a sector from arc $\widehat{AB}$ is given by $\frac{\theta}{360^\circ}\pi r^2$ where θ is the angular size of $\widehat{AB}$ (measured in degrees).

Theorems about Perpendiculars (Results will be proven below.)

- In a circle, a radius is perpendicular to a chord if and only if the radius bisects the chord.
- In a circle, the perpendicular bisector of a chord passes through the center of the chord.
- Similar to the above: Suppose a line going through a point P on the circle. The line is tangent to the circle if and only if the line is perpendicular to the radius of the circle.

Arcs and Angles (Results will be proven below.)

- If points A, B, P are on a circle, we call $\angle APB$ an *inscribed angle*. The measure of $\angle APB$ is half the angular size of arc $\widehat{AB}$ (where the arc does NOT contain P).
- Suppose two chords AC, BD intersect inside the circle at a point P as in the diagram below.

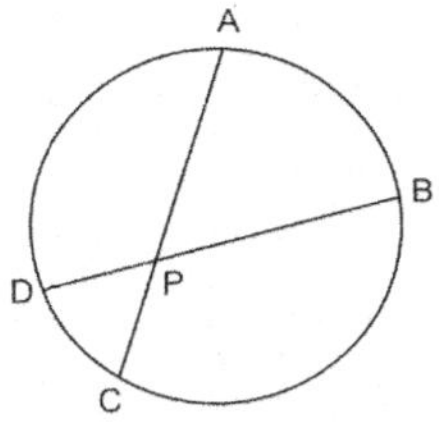

Then $\angle APB$ is half the sum of the angular sizes of arcs $\widehat{AB}$ and $\widehat{CD}$.
- Suppose the extension of two chords AC, BD intersect outside the circle at a point P as in the diagram below.

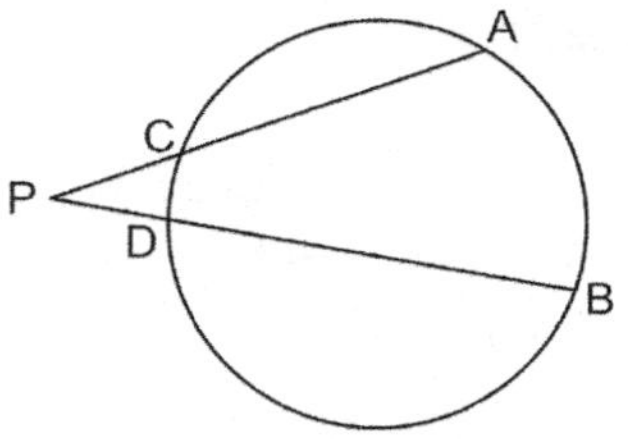

Then $\angle APB$ is half the difference of the angular sizes of arcs $\widehat{AB}$ and $\widehat{CD}$.

7.1 Example Questions

Problem 7.1 Storing Tires

(a) Suppose two tires, each with radius 1ft rest upright on the ground and touching each other, as pictured below:

How much space is needed horizontally to store the tires?

(b) Repeat part (a) with two tires of radius $1,2$ feet respectively.

Problem 7.2 Suppose you start with a circle of radius 1. Draw another circle of radius 1 with center an arbitrary point on the first circle. Let R denote the region consisting of all points that are inside both circles.

(a) Find the perimeter of R.

(b) Find the area of R.

Problem 7.3 Inscribing Circles in Sectors

(a) What is the radius of the largest circle that can fit in a quarter circle of radius 1?

(b) What if instead you are fitting it into a 60°-sector?

Problem 7.4 Let $\angle APB$ be an inscribed angle on a circle with center O. Prove that $\angle APB$ is half the angular size of arc $\widehat{AB}$ if:

(a) O lies on $\angle APB$.

(b) O lies inside $\angle APB$.

Problem 7.5 Prove that if two chords AC, BD intersect inside a circle at point P then the measure of $\angle APB$ is half the sum of the angular sizes of $\widehat{AB}, \widehat{CD}$.

Problem 7.6 Suppose $\widehat{AB}$ is an arc with angular size 60° and CD is a diameter such that if rays $\overrightarrow{BA}, \overrightarrow{DC}$ are extended to intersect at a point E, $\angle AEC = 30$. Find the angular size of arc $\widehat{BD}$.

Problem 7.7 Suppose two perpendicular chords intersect and divide each other in a ratio of $1:2$. Find the radius of the circle if each chord is 12in long.

Problem 7.8 Suppose ω is a circle with radius 6 and center O. Let $\widehat{AB} = 135^\circ$. Let C be on ω such that $\overline{OA} \parallel \overline{BC}$. Find $[OACB]$.

Problem 7.9 (AMC 12A 2007 #10) A triangle with side length in the ratio $3:4:5$ is inscribed in a circle of radius 3. What is the area of the triangle?

Problem 7.10 Let $\mathscr{C}_1$ and $\mathscr{C}_2$ be circles defined by

$$(x-10)^2+y^2=36$$

and

$$(x+15)^2+y^2=81,$$

respectively. What is the length of the shortest line segment $\overline{PQ}$ that is tangent to $\mathscr{C}_1$ at P and to $\mathscr{C}_2$ at Q?

7.2 Quick Response Questions

Problem 7.11 The circle $x^2+y^2+10x-24y-87=0$ has center (h,k). What is $h+k$?

Problem 7.12 What is the radius of the circle $x^2+y^2+10x-24y-87=0$?

Problem 7.13 If four circles are drawn in a plane, what is the maximum number of points that belong to at least two of the circles?

Problem 7.14 Arrange 4 congruent circles so that (i) the center of the four circles form a square with side length 10, and (ii) adjacent circles are tangent. What is the radius of each circle?

Problem 7.15 Arrange 4 congruent circles so that (i) the center of the four circles form a square with side length 10, and (ii) adjacent circles are tangent. Find the area of the region inside the square that is outside each of the circles. Use $\pi = 3.14$ and round your answer to the nearest tenth if necessary.

Problem 7.16 Suppose A,B,C are points on a circle such that the angular measures of arc $\widehat{AB}$ (not containing C) and arc $\widehat{CA}$ (not containing B) are in ratio $5:9$. Suppose further that $\angle ABC = 90°$. Find the measure of $\angle BAC$.

Problem 7.17 Suppose ω is a circle with radius 6 and center O. Let A, B and C be points on ω such that $\overline{OA} \parallel \overline{BC}$. What angle does $\widehat{AB}$ have to be so that $OACB$ is a parallelogram?

Problem 7.18 In the following diagram $\angle ABC = 42^\circ$. What is the measure of $\angle AOC$?

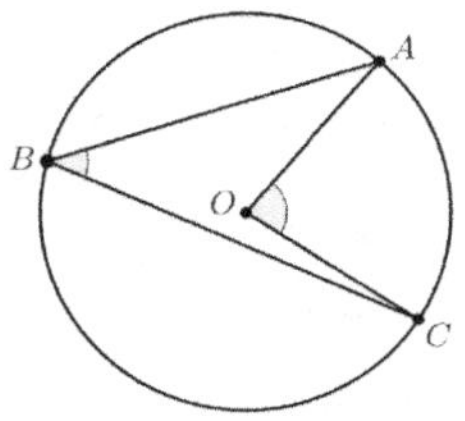

Problem 7.19 In the following diagram $\angle AOD = 100^\circ$ and $\angle BPC = 66^\circ$. What is the measure of angle $\angle BOC$?

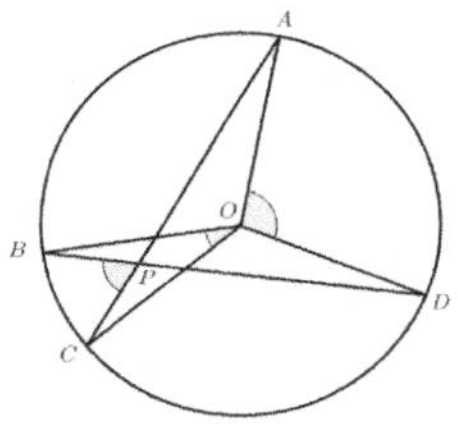

Problem 7.20 In the following diagram $\angle AOD = 115°$ and $\angle BOC = 23°$. What is the measure of angle $\angle BPC$?

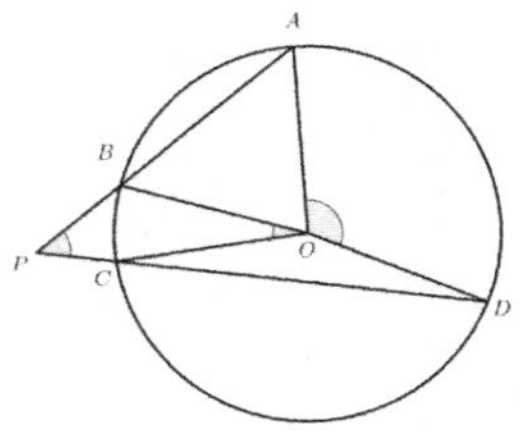

7.3 Practice Questions

Problem 7.21 Prove that in a circle, a radius is perpendicular to a chord if and only if the radius bisects the chord.

Problem 7.22 Prove that in a circle, the perpendicular bisector of a chord passes through the center of the circle.

Problem 7.23 Suppose you have three tires with radii $1, 2, 3$ feet respectively. You store them horizontally as in the example question. What is the minimum amount of horizontal space needed to store the tires? Justify your answer!

Problem 7.24 Suppose you start with a circle of radius 1. Draw a second circle of radius 1 centered at an arbitrary point on the first circle. Now pick one of the two intersection points and draw a third circle centered at this point, also with radius 1. Call R the region consisting of all points that are contained inside at least two circles.

(a) Find the perimeter of R.

(b) Find the area of R.

Problem 7.25 Inscribing Circles in Sectors

(a) What is the radius of the largest circle that can fit in a $120°$ sector of a circle of radius 1?

(b) What if you have a 240° sector?

Problem 7.26 Let $\angle APB$ be an inscribed angle on a circle with center O. Prove that $\angle APB$ is half the angular measure of arc $\widehat{AB}$ if O lies outside $\angle APB$.

Problem 7.27 Prove that if two chords AC, BD intersect outside a circle at point P then the measure of $\angle APB$ is half the difference of the angular sizes of $\widehat{AB}, \widehat{CD}$.

Problem 7.28 Suppose $\overline{AB}, \overline{CD}$ are two chords of equal length who intersect at E. Suppose $\angle AED = 120°$, and $AE : EB = CE : ED = 1 : 2$. Further, suppose $AC = 2$.

(a) Find the distance from E to the center of the circle.

(b) Find the radius of the circle.

Problem 7.29 Find the area of a 30-60-90 triangle circumscribed about a circle of radius 1.

Problem 7.30 Let $\mathscr{C}_1$ and $\mathscr{C}_2$ be circles defined by

$$(x-5)^2 + y^2 = 9$$

and

$$(x+5)^2 + y^2 = 9,$$

respectively. What is the length of the shortest line segment $\overline{PQ}$ that is tangent to $\mathscr{C}_1$ at P and to $\mathscr{C}_2$ at Q? Hint: There are a couple possible tangents.

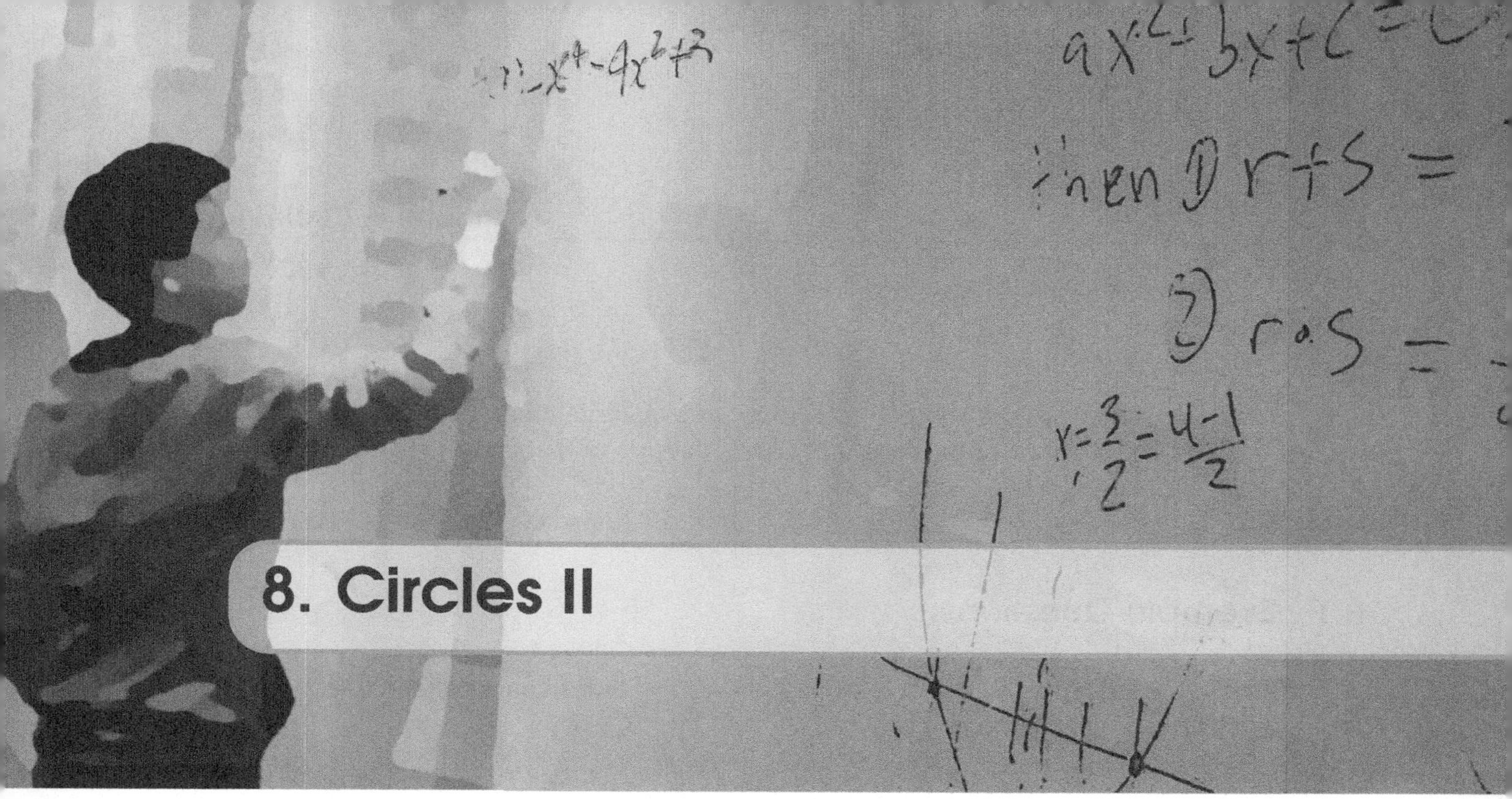

8. Circles II

Cyclic Quadrilateral (Results Proven Below)

- Points A, B, C, D lie on circle ω in (clockwise) order. (We say quadrilateral $ABCD$ is *cyclic*.) Then $\angle A + \angle C = \angle B + \angle D = 180^\circ$.
- Let $ABCD$ be a convex cyclic quadrilateral. Then $\angle DBA = \angle DCA, \angle ACB = \angle ADB, \angle BDC = \angle BAC$, and $\angle CBD = \angle CAD$.

Power of a Point: Special Cases

- Let $\overline{AB}, \overline{CD}$ be chords, which intersect at E inside the circle. Then $AE \cdot BE = CE \cdot DE$.
- Let $\overline{AB}, \overline{CD}$ be chords, which are extended to intersect at E outside the circle. Then $AE \cdot BE = CE \cdot DE$.
- Let a line tangent to the circle at C intersect the extension of chord $\overline{AB}$ at E. Then $AE \cdot BE = CE^2$. (Note: Think of this as the previous case with $C = D$.)

Power of a Point: Stronger Version (Results Proven Below)

- Let ω be a circle with center O and radius R, and let X be a point in the plane that is different from O. A line is drawn through X; it intersects ω at Y and Z. (If the line is tangent to the circle, then $Y = Z$). Then the product $XY \cdot XZ = XO^2 - R^2$ is constant (it does not depend on the line drawn), and it is called the power of X with respect to ω. (The lengths are directed.)

Ptolemy's Theorem (Results Proven Below)

- In a cyclic quadrilateral $ABCD$,

$$AC \cdot BD = AB \cdot CD + AD \cdot BC.$$

(In other words, if all the four vertices of a quadrilateral are on the same circle, then the product of the diagonals equals the sum of the products of the two pairs of opposite sides.)

8.1 Example Questions

Problem 8.1 Prove that if $\angle A + \angle C = \angle B + \angle D = 180^\circ$ in quadrilateral $ABCD$, then $ABCD$ is a cyclic quadrilateral .

Problem 8.2 Suppose $ABCD$ is a cyclic quadrilateral.

(a) Prove that $\angle ABD = \angle ACD$ (and similarly $\angle BAC = \angle BDC$, etc.).

(b) Prove that if E is the intersection of $\overline{AC}, \overline{BD}$ then $AE \cdot EC = BE \cdot ED$.

Problem 8.3 Power of a Point Inside a Circle

(a) Suppose $\overline{YZ}$ is a chord, with X the midpoint. Prove the Power of a Point formula for X and the line through Y, Z.

(b) Prove the Power of a Point formula for arbitrary X inside the circle.

Problem 8.4 Prove Ptolemy's theorem, using the following diagram as guidance,

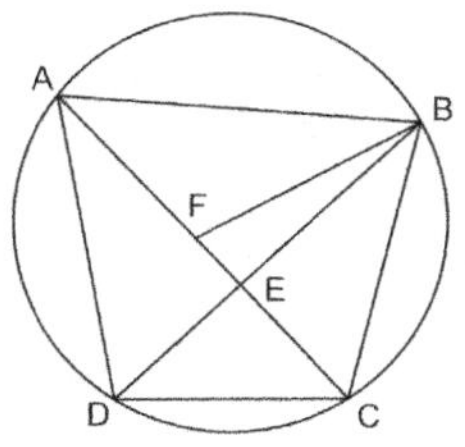

where F is such that $\angle ABF = \angle CBD$.

Problem 8.5 Suppose chords $\overline{AB}, \overline{CD}$ intersect at E, such that $AE : EB = 1 : 3$ and $CE : ED = 1 : 12$. Find the ratio of $AB : CD$.

Problem 8.6 (AHSME 1999 #21) A circle is circumscribed about a triangle with sides 20, 21, and 29, thus dividing the interior of the circle into four regions. Let A, B, and C denote the areas of the non-triangular regions, with C being the largest. Compute $C - (A + B)$.

Problem 8.7 Let A, B, C, D be four points, arranged in clockwise order, on circle ω. Segments AC and BD intersect at P. Given that $AB = 3, BP = 4, PA = 5, PC = 6$, find the radius of circle ω.

Problem 8.8 Suppose we have a rectangle $ABCD$ with $AB = 8$, $BC = 12$. Inscribe a circle in the rectangle so that it touches sides $\overline{AB}, \overline{BC}, \overline{AD}$. Let M be the midpoint of $\overline{AB}$. Call $E \neq M$ the intersection of $\overline{MD}$ with the circle. Find DE.

Problem 8.9 Given right triangle ABC, construct semicircles on the three sides as shown.

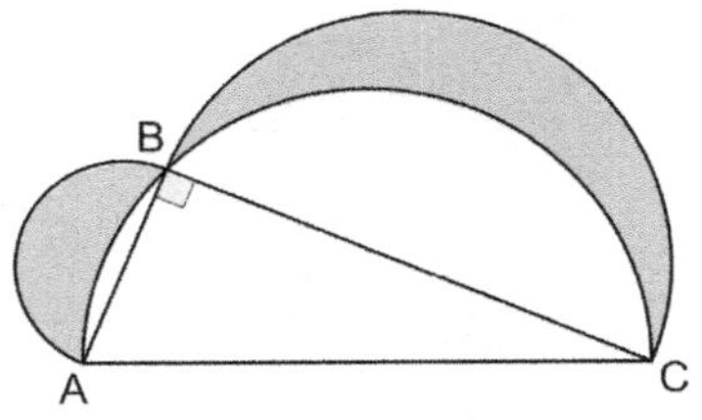

Given that $AB = 5, BC = 12$. Find the sum of the areas of the shaded regions.

Problem 8.10 In $\triangle ABC$, $AB = 37, AC = 58$. Use A as center and AB as radius, draw a circle to intersect $\overline{BC}$ at D where D is between B and C.

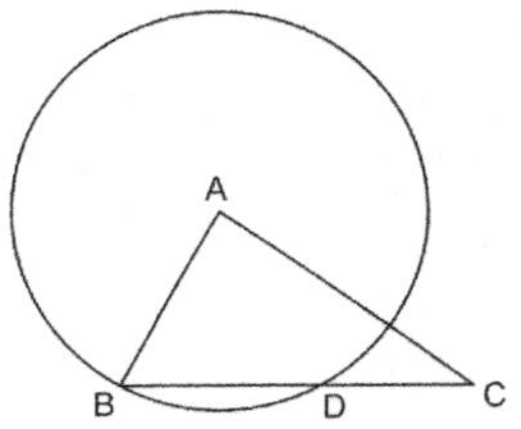

Given that the lengths of $\overline{BD}$ and $\overline{DC}$ are both integers, compute BC.

8.2 Quick Response Questions

Problem 8.11 Let $ABCD$ be a cyclic quadrilateral. Suppose $\angle DAB = 120^\circ$ and $\angle BDC = 40^\circ$, what is the measure of $\angle DBC$?

Problem 8.12 What is a better name for a cyclic parallelogram?

(A) Trapezoid
(B) Rhombus
(C) Rectangle
(D) None of the above

Problem 8.13 Let $ABCD$ be a quadrilateral with $\angle DAB = 116^\circ$, $\angle ABC = 92^\circ$, $\angle BCD = 60^\circ$ and $\angle CDA = 92^\circ$. Is it true that $ABCD$ is a cyclic quadrilateral?

Problem 8.14 Let $ABCD$ be a quadrilateral with $\angle DAB = 87^\circ$, $\angle ABC = 57^\circ$, $\angle BCD = 93^\circ$ and $\angle CDA = 123^\circ$. Is it true that $ABCD$ is a cyclic quadrilateral?

Problem 8.15 Suppose a trapezoid $ABCD$ with base $\overline{AD}$ is inscribed in a circle. If $AB = 4$, what is CD?

Problem 8.16 Let $\overline{PQ}$ be a chord with $PQ = 12$. Extend $\overline{PQ}$ to a point R such that $QR = 4$. Let T be such that $\overline{RT}$ is tangent to the circle. What is RT?

Problem 8.17 Let A, B, C and D be points on circle $\mathscr{C}$ such that AB and CD intersect at point P outside of $\mathscr{C}$. If $AB = 10$, $AP = 16$, and $CP = 15$, what is DP?

Problem 8.18 Let A, B, C and D be points on a circle $\mathscr{C}$ such that AB and CD intersect at point P inside of $\mathscr{C}$. If $AB = 10$, $AP = 4$, and $CP = 3$, what is DC?

Problem 8.19 Suppose O is the center of a unit circle. Let $ABCO$ be a rhombus with A, B, C on the circle. $[ABCO] = \frac{\sqrt{P}}{Q}$. What is $P + Q$?

Problem 8.20 Suppose equilateral triangle ABC is inscribed in a circle. Let P be on minor arc $\widehat{AB}$ such that $AP = 1$ and $BP = 3$ (distances of the line segments). Find the distance PC.

8.3 Practice Questions

Problem 8.21 Prove the Pythagorean theorem using Ptolemy's theorem.

Problem 8.22 Prove that if $ABCD$ is a cyclic quadrilateral, then $\angle A + \angle C = \angle B + \angle D = 180°$.

Problem 8.23 Suppose $ABCD$ is a quadrilateral.

(a) Prove that if $\angle ABD = \angle ACD$ (or $\angle BAC = \angle BDC$, etc.) then $ABCD$ is cyclic.

(b) Let $\overline{AC}, \overline{BD}$ intersect at E. If $AE \cdot EC = BE \cdot ED$, show that $ABCD$ is cyclic.

Problem 8.24 Prove the Power of a Point Formula for X outside the circle. Hint: First consider a line that is tangent to the circle. Then proceed similar to the case for inside the circle.

Problem 8.25 Suppose diameter $\overline{CD}$ intersects chord $\overline{AB}$ at E, so that $AE = 4, EB = 9$. If the diameter of the circle is 15, Find CE and ED.

Problem 8.26 Suppose you have a circle with area 1. Inscribe a triangle in the circle. Let A, B, and C denote the area of the non-triangular regions, with C being the largest.

(a) Show that $C - (A + B)$ can get very close to 1.

(b) Find the minimum value of $C-(A+B)$. Hint: It is negative!

Problem 8.27 Suppose $ABCD$ is an isosceles trapezoid (so it is cyclic). Further suppose the center of this circle is in the interior of the trapezoid $ABCD$, and the radius is 25, with $AD=40$ and $BC=48$. Find the area of trapezoid $ABCD$.

Problem 8.28 Suppose we have a rectangle $ABCD$ with $AB=8$, $BC=12$. Inscribe a circle in the rectangle so that it touches sides $\overline{AB},\overline{BC},\overline{AD}$. Let M be the midpoint of $\overline{AB}$. Call $E\neq M$ the intersection of $\overline{MD}$ with the circle. Assume that CE is tangent to the circle at E (It is, as a challenge try to prove it!). Find the length of CE.

Problem 8.29 Inscribe equilateral $\triangle ABC$ in a circle of radius 1. Then construct semicircles on the three sides as shown.

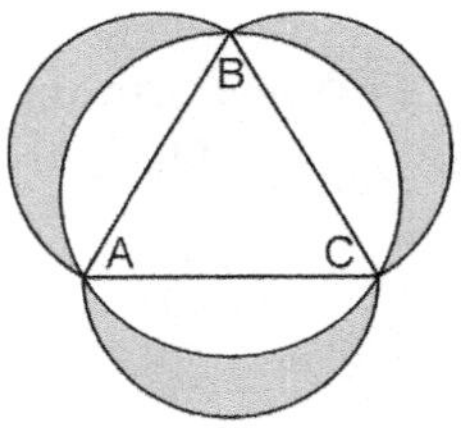

Find the sum of the areas of the shaded regions.

Problem 8.30 Let O be the circumcenter of $\triangle ABC$. Through A construct a line tangent to $\odot O$ and intersecting the extension of $\overline{BC}$ at D. From B and C construct lines perpendicular to $\overline{AD}$ with feet M and N respectively. Assume that $CN=5, BC=5, AD=5\sqrt{6}$. Find the area of trapezoid $MBCN$.

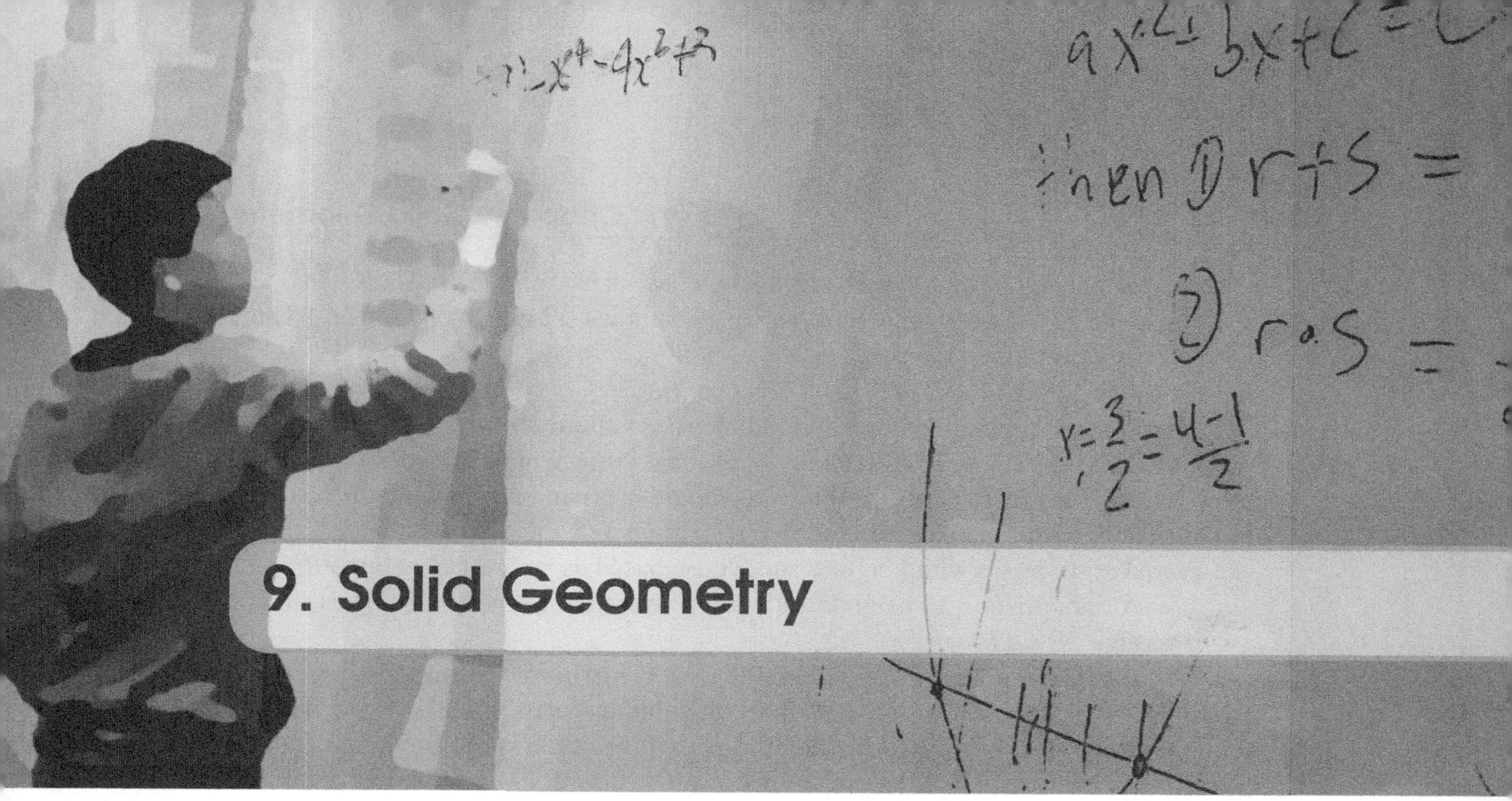

9. Solid Geometry

Basics

- Points in three dimensions can be thought of as using x, y, z coordinates, written (x, y, z).
- We have analogues of the Pythagorean Theorem and Distance Formula in three dimensions: The distance from $(0,0,0)$ to (a,b,c) is $\sqrt{a^2+b^2+c^2}$.
- Given two distinct lines, there are three possibilities: (i) they are parallel, (ii) they intersect, or (iii) they are *skew*.
- Given two distinct planes, there are two possibilities: (i) they are parallel, or (ii) they intersect (and their intersection is a line).

Solids

- **Sphere**: The collection of points (in three dimensions) of equal distance (called the *radius* from a *center*).
  - A sphere with radius r has volume $\frac{4}{3}\pi r^3$.
  - A sphere with radius r has surface area $4\pi r^2$.
- **Cube**: The 3-D version of a square.
  - A cube with side length s has volume s^3.
  - A cube with side length s has surface area $6s^2$.
  - A cube has 8 vertices, 12 edges, and 6 faces. The 6 faces are all squares.
- **Rectangular Prism**: A "box".
  - A rectangular prism with sides l, w, h (often 'length', 'width', 'height') has volume lwh.

- A rectangular prism with sides l, w, h has surface area $2(lw + wh + lh)$.
- Like a cube, a rectangular prism has 8 vertices, 12 edges, and 6 faces. The 6 faces are all rectangles.

- **Cylinder**: A "can".
  - A cylinder with height h and radius r has volume $\pi r^2 h$.
  - A cylinder with height h and radius r has surface area $2\pi r^2 + 2\pi rh$.
- **(General) Right Prism**: Similar to rectangular prism or cylinders with an arbitrary polygon as the base (the 'top' and 'bottom').
  - A right prism with height h and whose base has area B has volume Bh.
  - A right prism with height h and base with area B and perimeter P has surface area $2B + Ph$.
  - If the base polygon has n sides, the right prism has $2n$ vertices, $3n$ edges, and $n+2$ faces. Every side face is rectangular and perpendicular to the bases.
  - **Note**: The volume formula holds even if the prism is not right, that is if the two bases are parallel but not necessarily 'above' each other.
- **Square Right Pyramid**: The standard "pyramid from Egypt" solid, with a square *base* and *apex* (or top point) that is centered above the square.
  - A square right pyramid with height h with square base of side length s has volume $\frac{1}{3}s^2 h$.
  - A square right pyramid with height h with square base of side length s has surface area $s^2 + 2sL$, where $L = \sqrt{h^2 + s^2/4}$ (L is called the *slant height*).
  - A square right pyramid has 5 vertices, 8 edges, and 5 faces. The 4 side faces are all triangles.
- **Right Cone**: The standard "ice cream cone" solid, with a circular *base* and *apex* that is centered above the circle.
  - A right with height h with square base of radius r has volume $\frac{1}{3}\pi r^2 h$.
  - A square right pyramid with height h with square base of side length s has surface area $\pi r^2 + \pi rL$, where $L = \sqrt{h^2 + r^2}$ (L is called the *lateral height*).
- **(General) Right Pyramid**: Similar to a square right pyramid or right cone with an arbitrary polygon as the base.
  - A right pyramid with height h and whose base has area B has volume $\frac{1}{3}Bh$.
  - A right pyramid with height h and base with area B and perimeter P has surface area $B + PL/2$, where $L = \sqrt{h^2 + r^2}$, where r is the inradius of the base (L is called the *slant height*).
  - If the base polygon has n sides, the right pyramid has $n+1$ vertices, $2n$ edges, and $n+1$ faces. The n side faces are all triangles.
  - **Note**: The volume formula holds even if the pyramid is not right, that is if the apex is not necessarily centered above the base.

- **Tetrahedron**: A triangular pyramid. Alternatively a solid made up of four triangles.
  - As a tetrahedron is a triangular pyramid, the above formulas hold.
  - In particular, a *regular* tetrahedron is a tetrahedron made up of 4 equilateral triangles.

9.1 Example Questions

Problem 9.1 Inscribing Spheres in a Cone

(a) Find the volume of the largest sphere that can fit inside a cone of radius 1 and height $\sqrt{3}$.

(b) Assume the sphere as in (a) is placed inside the cone. Suppose you now want to fit another sphere in the cone that is tangent to the base. Find the radius of the largest such sphere.

Problem 9.2 Suppose you have a regular tetrahedron with side length a.

(a) Show that the surface area is $a^2\sqrt{3}$.

(b) Show that the volume is $\dfrac{a^3\sqrt{2}}{12}$.

Problem 9.3 (2010 AMC 10A #20) Suppose a bored bee lives on a cube with side length 1. For "fun" he decides to visit every vertex of the cube, each exactly once, starting and ending at the same vertex. It will travel from one vertex to another using straight lines (either crawling or flying). Give an example of a path that uses the maximum distance and find this distance.

Problem 9.4 Four identical balls (spheres), each of radius 1in, are glued to the ground so that their centers form the vertices of a square with side length 2in. Suppose you rest a fifth identical ball on the four balls (so the fifth ball is a sphere externally tangent to the other spheres). How far does this ball rest off the ground?

Problem 9.5 Suppose you pick 4 vertices of a cube to form a tetrahedron.

(a) How many different (non-congruent) tetrahedra are possible? Are any of them regular tetrahedra?

(b) Find the volumes for each of the possibilities in (a) if the cube has volume 1.

Problem 9.6 Suppose you have a regular square pyramid $S-ABCD$ with height 6 whose square has side length 4. Call the midpoints of the square E,F,G,H (on $\overline{AB},\overline{BC},\overline{CD},\overline{DA}$ respectively) and the midpoints of $\overline{SA},\overline{SB},\overline{SC},\overline{SD}$ respectively T,U,V,W. Form polyhedron $EFGH-TUVW$ (with 10 faces).

(a) Describe the faces of $EFGH-TUVW$. How many vertices and edges does the polyhedron have?

(b) Find the volume of $EFGH-TUVW$. Hint: Do this indirectly.

Problem 9.7 Suppose you have a unit cube. Pick two opposite corners. In each corner, form a tetrahedron using the corner and the three adjacent vertices. Remove these two tetrahedra and call the resulting polyhedron $\mathscr{S}$.

(a) How many vertices, edges, and faces does the resulting polyhedron have? Describe the faces.

(b) Find the volume of $\mathscr{S}$.

Problem 9.8 Suppose you have tetrahedron $S-ABC$. Cut the tetrahedron with a plane between S and $\triangle ABC$, forming a new tetrahedron $S-A'B'C'$ (with A' on $\overline{SA}$, etc.). Prove that the ratio of volumes of $S-ABC$ to $S-A'B'C'$ is equal to $SA \cdot SB \cdot SC : SA' \cdot SB' \cdot SC'$.

Problem 9.9 (2012 AMC 10A #21) Let points $A=(0,0,0)$, $B=(3,0,0)$, $C=(0,4,0)$, $D=(0,0,5)$. Suppose E,F,G,H are midpoints of (respectively) $\overline{BD},\overline{AB},\overline{AC},\overline{DC}$. Prove that $EFGH$ is a square.

Problem 9.10 Suppose you start with a right cone and cut off the top of the cone with a plane parallel to the base. The resulting solid, called a *frustrum* has two circular "bases", say with radii R and r (with $R>r$), and height h. (Hence from the side the frustrum looks like a trapezoid with bases R,r and height H.)

(a) Show that the volume of a frustrum is $\dfrac{\pi H}{3}(R^2+Rr+r^2)$.

(b) Suppose a sphere can be inscribed in a frustrum with base radii r, R such that the sphere is tangent to the two bases and the side. Find the radius of such a sphere in terms of r, R.

9.2 Quick Response Questions

Problem 9.11 How many total pairs of parallel edges does a cube contain?

Problem 9.12 Suppose we have a ball with radius 6. Suppose you cut the ball in half. What is the volume of the half-ball? Use $\pi = 3.14$ and round your answer to the nearest tenth if necessary.

Problem 9.13 Suppose we have a ball with radius 6. Suppose you cut the ball in half. What is the surface area of the half-ball? Use $\pi = 3.14$ and round your answer to the nearest tenth if necessary.

Problem 9.14 A cube is increased to form a new cube so that the surface area of the new cube is 64 times that of the original cube. By what factor is the volume of the cube increased?

Problem 9.15 A cube is increased to form a new cube so that the volume of the new cube is 64 times that of the original cube. By what factor is the surface area of the cube increased?

Problem 9.16 In the $2 \times 2 \times 2$ cubic figure below, if the path is required to be along the surface of the cube, what is the length of the shortest path from point A to B. This length can be expressed in the form $\sqrt{K}$ for an integer K. What is K?

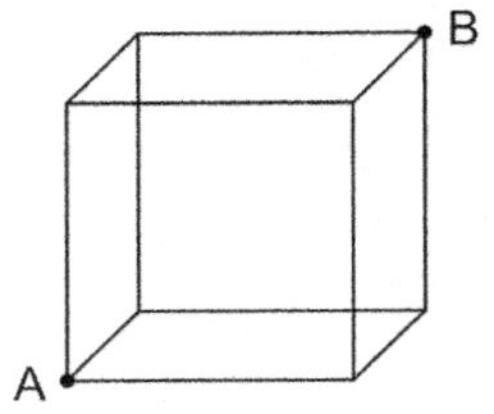

Problem 9.17 A regular tetrahedron has volume $\frac{8}{3}$. What is the side length, rounded to the nearest tenth if necessary?

Problem 9.18 What is the surface area of a square pyramid with base side length 6 and height 4? Round your answer to the nearest integer if necessary.

Problem 9.19 Suppose you have an ice cream cone with radius 2 inches and height 4 inches. The cone starts full of ice cream (but there is no ice cream outside the cone). After you've eaten some ice cream and some of the cone you are left with a cone with a radius and a height of 2 inches. What percent of the ice cream have you eaten?

Problem 9.20 An obtuse triangle with dimensions 9, 10, and 17 is rotated about the smallest side so that it creates a three-dimensional solid shown below.

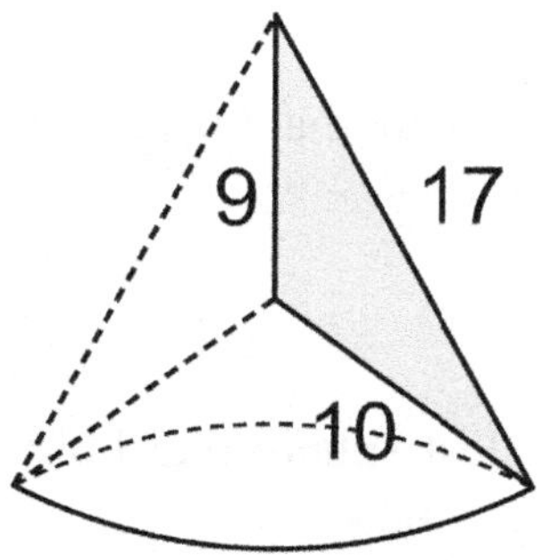

Determine the surface area of the solid. Use $\pi = 3.14$ and round your answer to the nearest tenth if necessary.

9.3 Practice Questions

Problem 9.21 Inscribing Spheres and Cubes

(a) Find the volume of the largest sphere that fits in a cube of volume 1. (That is, inscribe a sphere inside the cube.)

(b) Find the volume of the smallest sphere that holds a cube of volume 1. (That is, circumscribe a sphere outside the cube.)

Problem 9.22 Four identical balls (spheres), each of radius 1in, are glued to the ground so that their centers form the vertices of a square with side length 2in. Suppose you rest a fifth ball that rests 1in off the ground. Find the radius of the fifth ball.

Problem 9.23 Suppose $S-ABC$ is a regular tetrahedron with apex S. Cut off the top half of the tetrahedron (that is, cut through the midpoints of $\overline{SA}, \overline{SB}, \overline{SC}$ and leave the bottom).

(a) How many vertices, edges, and faces does the resulting solid have?

(b) Find the volume and the surface area as ratios to the original volume and surface area.

Problem 9.24 Suppose a bored bee lives on a cube with side length 1. For fun he decides to visit every vertex of the cube, each exactly once, starting and ending at the same vertex. It will travel from one vertex to another using straight lines (either crawling or flying).

(a) What is the distance of the shortest such path?

(b) Note in part (a) the shortest path was on the surface of the cube (that is, did not travel inside the cube). What is the longest path if the bee is not allowed to travel inside the cube?

Problem 9.25 Suppose you have a sphere of radius 1. That is the side length of the largest regular tetrahedron you can fit (inscribe) inside the sphere?

Problem 9.26 Let $S-ABC$ be a regular tetrahedron with volume 1. Let E,F,G be the midpoints of $\overline{AB},\overline{BC},\overline{AC}$ and J,K,L be the midpoints of $\overline{SA},\overline{SB},\overline{SC}$. Form solid $EFG-JKL$. What is the volume of this solid?

Problem 9.27 As in class, form a solid by removing opposite tetrahedra in a unit cube. (Here each tetrahderon is formed by a vertex and the three adjacent vertices in a cube.) Suppose $\mathscr{S}$ is resting on one of the faces (ignore whether the polyhedron would actually balance or not). What are different possible heights of $\mathscr{S}$?

Problem 9.28 Start with right triangular prism $ABC-DEF$. Divide the prism into four parts using the planes through points A,B,F and D,E,C. Find the ratio of the volumes of these four parts.

Problem 9.29 Let points $A=(0,0,0)$, $B=(3,0,0)$, $C=(0,4,0)$, $D=(0,0,5)$. Suppose E,F,G,H are midpoints of (respectively) $\overline{BD},\overline{AB},\overline{AC},\overline{DC}$ (recall in Problem 9.9 we have shown that $EFGH$ is a square). Cut tetrahedron $D-ABC$ into two pieces using square $EFGH$. Find the volumes of each of these pieces.

Problem 9.30 Recall in Problem 9.10 we discussed frustrums and when it was possible to inscribe a sphere in it. A sphere is inscribed in a frustrum with bases of radii 1 and 2. The intersection of this sphere with the side of the frustrum is a circle. Find the radius of this circle.

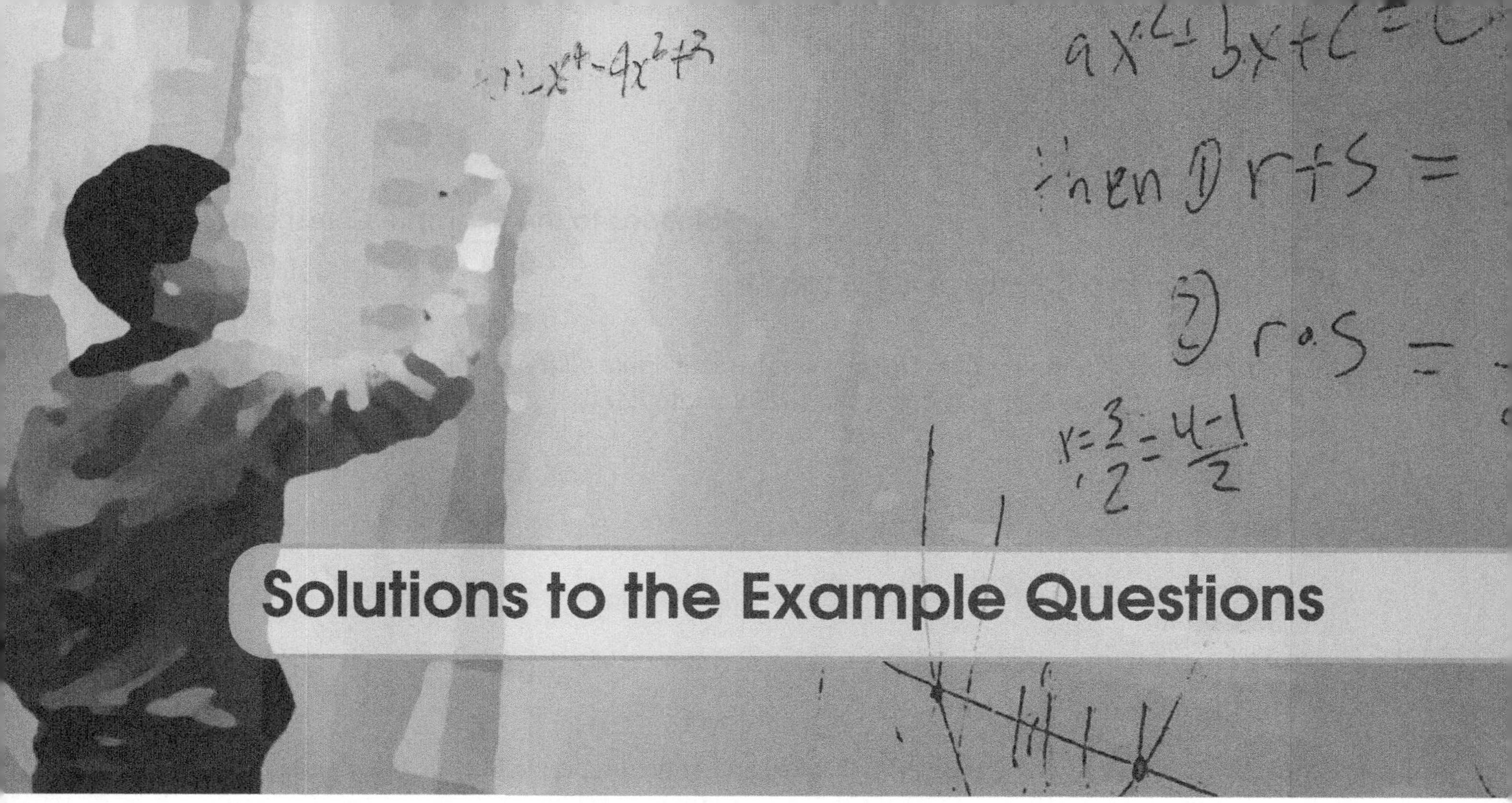

Solutions to the Example Questions

In the sections below you will find solutions to all of the Example Questions contained in this book.

Quick Response and Practice questions are meant to be used for homework, so their answers and solutions are not included. Teachers or math coaches may contact Areteem at info@areteem.org for answer keys and options for purchasing a Teachers' Edition of the course.

1 Solutions to Chapter 1 Examples

Problem 1.1 Suppose that $ABCD$ is a square, and that CDP is an equilateral triangle, with P *outside* the square. What is the size of angle PAD?

Answer

15°.

Solution

We have $\angle ADP = \angle ADC + \angle CDP = 90 + 60 = 150^\circ$. Since $\triangle ADP$ is isosceles $\angle PAD = (180 - 150)/2 = 15^\circ$.

Problem 1.2 Given square $ABCD$, let P and Q be the points outside the square that make triangles CDP and BCQ equilateral. Segments AQ and BP intersect at T. Find angle ATP.

Answer

90°.

Solution

$\triangle ABQ \cong \triangle BCP$ (why?), therefore $\angle BAT = \angle TBC$ (equals 15° by a previous problem). Consider $\triangle ABT$. Since $\angle BAT = \angle TBC$, $\angle BAT + \angle ABT = 90^\circ$ and thus $\angle ATB = 90^\circ$. This implies $\angle ATP = 90^\circ$ as well.

Problem 1.3 Squares $OPAL$ and $KEPT$ are attached to the outside of equilateral triangle PEA. Draw segment TO, then find the size of angle TOP.

Answer

30°.

Solution

We have $\angle TPO = 360 - 90 - 60 - 90 = 120^\circ$. As $\triangle TOP$ is isosceles $\angle TOP = (180 - 120)/2 = 30^\circ$.

Problem 1.4 Mark P inside square $ABCD$, so that triangle ABP is equilateral. Let

Q be the intersection of BP with diagonal AC. Triangle CPQ looks isosceles. Is this actually true?

Answer

Yes.

Solution

Using isosceles triangles, it is easy to see that $\angle BPC = \angle BCP = 75^\circ$. Since $\angle BCA = 45^\circ$, we get $\angle PCQ = 75^\circ - 45^\circ = 30^\circ$. Hence $\angle PQC = 75^\circ = \angle CPQ$. So $\triangle CPQ$ is a 30°-75°-75° triangle.

Problem 1.5 Let triangle ABC be equilateral triangle with side length 16. Let D be on side $\overline{AB}$ and E be on side $\overline{AC}$ such that $\overline{DE} || \overline{BC}$. Assume triangle ADE and trapezoid $DECB$ have the same perimeter. What is the length of $\overline{AD}$?

Answer

12.

Solution

Let $x = AD$. Note that $\triangle ADE$ is equilateral (why?). Thus the perimeter of $\triangle ADE = 3x$ and the perimeter of trapezoid $DECB$ is $x + 2(16 - x) + 16 = 48 - x$. Since these two perimeters are equal, $3x = 48 - x$ so $x = 12$.

Problem 1.6 Let E be a point inside unit square $ABCD$ such that CDE is an equilateral triangle. Find the area of triangle AEC.

Answer

$(\sqrt{3} - 1)/4$.

Solution

Note that $[AEC] = [AED] + [CDE] - [ADC]$. Further, (why?) $[AED] = \frac{1}{2} \cdot 1 \cdot \frac{1}{2}, CDE = \frac{1}{2} \cdot 1 \cdot \frac{\sqrt{3}}{2} = \frac{\sqrt{3}}{4}, [ADC] = \frac{1}{2} \cdot 1 \cdot 1$. Hence, $[AEC] = \frac{1}{4} + \frac{\sqrt{3}}{4} - \frac{1}{2} = \frac{\sqrt{3} - 1}{4}$.

Problem 1.7 A triangle has a 60-degree angle and a 45-degree angle, and the side opposite the 45-degree angle has length 12. How long is the side opposite the 60-degree angle?

Answer

$6\sqrt{6}$.

Solution

It is easy using the Law of Sines. Solution without the Law of Sines is as follows. The remaining angle is 75°. Draw the altitude from the vertex of 75°, and this altitude has length $6\sqrt{3}$ based on the 30-60-90 triangle, and then the required side length is $6\sqrt{6}$ based on the 45-45-90 triangle.

Problem 1.8 Let ABC be a triangle with $AB = AC$ and $\angle BAC = 20^\circ$, and let P be a point on side AB such that $AP = BC$. Construct point D such that triangle ACD is equilateral, as shown in the diagram below. Show that triangle DCP is isosceles.

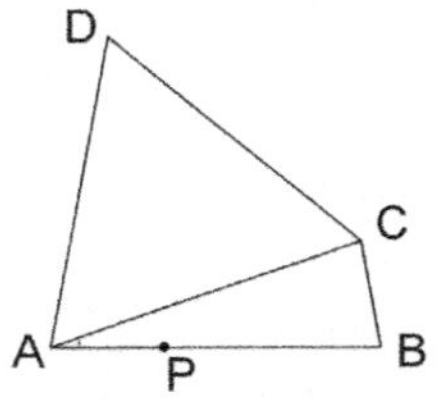

Solution

Connect $\overline{PD}$. Then $\triangle ABC \cong \triangle DAP$ by SAS congruency (double check this!). Thus $DP = AC = DC$ and thus $\triangle DCP$ is isosceles.

Problem 1.9 In triangle ABC, $AB = AC$, and BD is the altitude on AC. Given that $BD = \sqrt{3}$, and $\angle DBC = 60^\circ$, find the area of $\triangle ABC$.

Answer

$\sqrt{3}$.

Solution

First note that since $\angle DBC = 60^\circ$, we must have the altitude BD intersects the extension of AC, as in the diagram below.

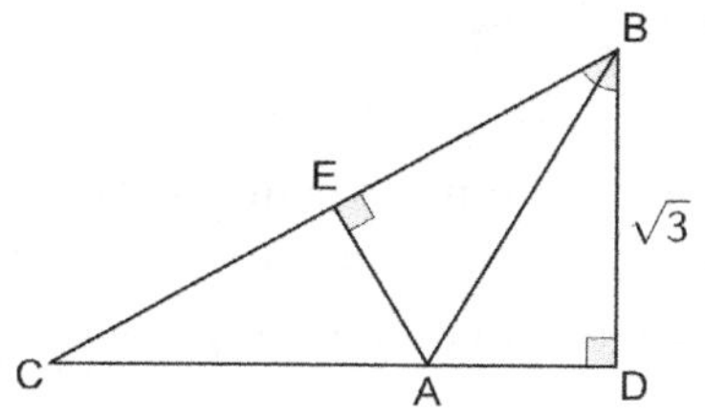

Since $\triangle BCD$ is a 30-60-90 triangle, we have $BC = 2\sqrt{3}$. Letting E be the midpoint of BC (so also AE is the altitude on BC) we have $\triangle AEC \cong \triangle AEB \cong \triangle ADB$ and all are 30-60-90 triangles themselves. Hence $AE = 1$ so the area of ABC is $\frac{1}{2} \cdot 2\sqrt{3} \cdot 1 = \sqrt{3}$ (using BC as the base).

Problem 1.10 In a right triangle $\triangle ABC$ suppose $\angle B = 90°$ and $\angle C = 30°$. Suppose point D is on $\overline{BC}$ with $\angle ADB = 45°$ and $DC = 10$. Find the length of AB.

Answer

$10/(\sqrt{3}-1)$.

Solution

Let $x = AB$. Then as $\triangle ABD$ is a 45-45-90 triangle, so $DB = x$ as well. As $\triangle ABC$ is a 30-60-90 triangle, $BC = \sqrt{3}AB$ so $x + 10 = x\sqrt{3}$. Hence $x = \dfrac{10}{\sqrt{3}-1}$.

2 Solutions to Chapter 2 Examples

Problem 2.1 Complete the following table about polygons with n sides: name, sum of interior angles, sum of exterior angles, and measure of each angle in case of regular polygon. All angles are in degrees. Justify your answers. Keep the chart for your own reference.

n	Name	Int. Angle Sum	Ext. Angle Sum	Each Angle (if regular)
3	Triangle			
4				
5				
6				
7	Heptagon			
8				
9	Nonagon			
10				
12	Dodecagon			
20	Icosagon			

Solution

To justify the sum of interior angles of a triangle: In $\triangle ABC$, draw line through A that's parallel to $\overline{BC}$, then use the property of alternate interior angles. For polygons of more sides: cut them into triangles, and get the formula $(n-2)180°$.

For sum of exterior angles: Pretend the polygon's sides are streets, you are walking along the streets. Each time you turn a corner, you turn an exterior angle; when you return to the starting point, you turned a total of 360°. You can also use the formula to calculate it. This gives the table below:

n	Name	Int. Angle Sum	Ext. Angle Sum	Each Angle (if regular)
3	Triangle	180°	360°	60°
4	Quadrilateral	360°	360°	90°
5	Pentagon	540°	360°	108°
6	Hexagon	720°	360°	120°
7	Heptagon	900°	360°	$900/7°$
8	Octagon	1080°	360°	135°
9	Nonagon	1260°	360°	140°
10	Decagon	1440°	360°	144°
12	Dodecagon	1800°	360°	150°
20	Icosagon	3240°	360°	162°

Problem 2.2 Use 6 equilateral triangles to form a hexagon $ABCDEF$.

(a) Show hexagon $ABCDEF$ is regular. Justify your answer.

Solution

It is clear that each side has the same length. Each angle in the hexagon is made up of two 60° angles (from the equilateral triangle), hence each angle is 120°.

(b) Calculate the angle AED.

Answer

90°

Solution

Note that $\triangle AFE$ is isosceles with $\angle AFE = 120°$. Thus, $\angle FEA = (180 - 120)/2 = 30°$ and $\angle AED = \angle FED - \angle FEA = 120 - 30 = 90°$.

Problem 2.3 Four non-overlapping regular plane polygons all have sides of length 1. The polygons meet at a point A in such a way that the sum of the four interior angles at A is 360°. Among the four polygons, two are squares and one is a triangle. What is the last polygon?

Answer

Hexagon

Solution

The other polygon's interior angle is 120°, so it is a hexagon.

Problem 2.4 Find the area of the largest equilateral triangle that fits in a regular hexagon of area 50.

Answer

25

Solution

If the hexagon is $ABCDEF$, the largest equilateral triangle is ACE (or equivalently BDF). If O is the center of the hexagon, note that $\triangle ABC \cong \triangle ACO$ (why?). Similar results hold for the rest of the hexagon, and from there it is not hard to show that $[ACE] = [ABCDEF]/2 = 25$.

Problem 2.5 In equiangular octagon $ABCDEFGH$, $AB = CD = EF = GH = 6\sqrt{2}$ and $BC = DE = FG = HA$. Given the area of the octagon is 184, compute the length of side BC.

Answer

4

Solution 1

Let $x = BC$, connect $\overline{AD}, \overline{EH}, \overline{BG}, \overline{CF}$. The sides $\overline{AB}, \overline{CD}, \overline{EF}, \overline{GH}$ are hypotenuses of four isosceles right triangles, whose legs are equal to 6. Thus the area of the whole octagon is

$$x^2 + 4 \cdot 6 \cdot x + 4 \cdot \frac{1}{2} \cdot 6 \cdot 6 = 184,$$

so

$$x^2 + 24x - 112 = 0.$$

We want the positive root: $x = 4$.

Solution 2

A better solution involves extending the sides $\overline{BC}, \overline{DE}, \overline{FG}, \overline{HA}$ to both directions to make a big square. The added corners of this square are isosceles right triangles with hypotenuses $6\sqrt{2}$, so their legs are all 6. Let $x = BC$, the big square's side length is $x + 12$. The added corners have a total area of two squares of side 6, so the added area is 72. Thus $(x+12)^2 = 184 + 72 = 256$, thus $x + 12 = 16$, and $x = 4$.

Problem 2.6 Answer the following

(a) Given a 60-120 isosceles trapezoid, prove that the sum of the length of the top and one of the side is equal to the length of the base.

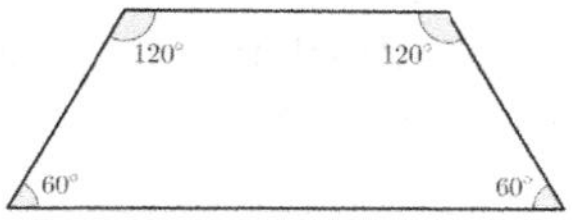

Solution

Divide the trapezoid as follows:

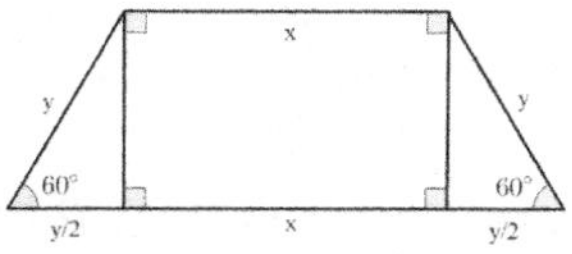

Let x denote the length of the top of the trapezoid and y denote the lengths of each side. If the trapezoid is divided into a rectangle and two 30-60-90 Triangles as shown, it is easy to see the length of the base of the trapezoid is equal to $y/2+x+y/2=x+y$.

(b) Let triangle ABC be a equilateral triangle with length equals to 1. Let point P be the center of the triangle ABC, D be point on $\overline{BC}$, E be point on $\overline{CA}$, and F be point on $\overline{AB}$, so that $\overline{PD}||\overline{AB}$, $\overline{PE}||\overline{BC}$, and $\overline{PF}||\overline{AC}$. Find the value of $PD+PE+PF$.

Answer

1

Solution

Note that $PDCE, PFBD, PEAF$ are all isosceles trapezoids. By the isosceles trapezoid problem, $PD+PE=CD$. Further, $PF=DB$, so $PD+PE+PF=CD+DB=CB=1$.

Problem 2.7 Show that a regular dodecagon (12-sided polygon) can be cut into pieces that are all regular polygons, which need not all have the same number of sides.

Solution

A regular hexagon at the center, surrounded by 6 squares and 6 equilateral triangles, all having the same side length.

Problem 2.8 What is the side length of the largest equilateral triangle that can fit inside a 2-by-2 square?

Answer

$2(\sqrt{6}-\sqrt{2})$.

Solution

In square $ABCD$, let the largest equilateral triangle be $\triangle AEF$, where E is on BC and F is on CD. Let $x=EC=CF$, then $BE=DF=2-x$. So $2^2+(2-x)^2=2x^2$, solving for x, we get $x=2(\sqrt{3}-1)$, and the side length of $\triangle AEF$ is $2(\sqrt{6}-\sqrt{2})$.

Problem 2.9 Let $ABCD$ be a unit square and let P, Q be on sides $\overline{AD}, \overline{AB}$ respectively such that $\triangle APQ$ has perimeter 2. Rotate $\triangle PDC$ 90° about C. Call the point P is rotated to P'. Prove that $\triangle PQC$ is congruent to $\triangle P'QC$.

Solution

Since $ABCD$ is a unit square, it is not hard to show that PQ has the same length as QP'. Then use SSS to show the triangles are congruent.

Problem 2.10 In convex quadrilateral $ABCD$, $\angle A = 60^\circ$, $\angle C = 30^\circ$, and $AB = AD$. Show that $AC^2 = BC^2 + CD^2$.

Solution

First recognize that $\triangle ABD$ is equilateral. Construct equilateral triangle ACC' where B is in the interior of $\triangle ACC'$ as in the diagram below.

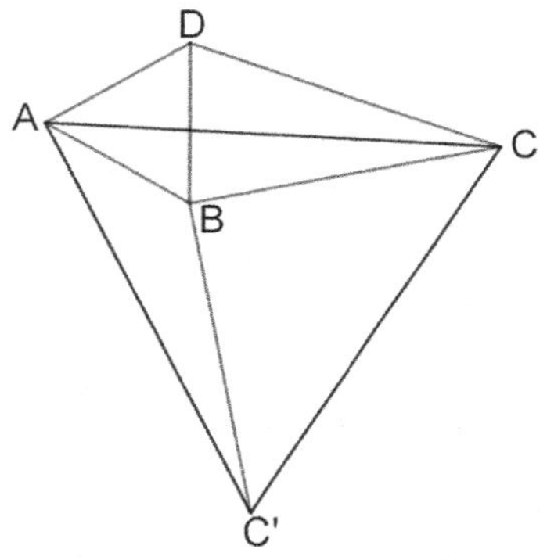

Since $AD = AB$, $\angle DAC = \angle BAC' = 60^\circ - \angle CAB$, and $AC = AC'$, we have $\triangle ADC \cong \triangle ABC'$. Thus $\angle ABC' = \angle ADC$. Since $\angle ADC + 60^\circ + \angle ABC + 30^\circ = 360^\circ$, and $\angle ABC' + \angle ABC + \angle CBC' = 360^\circ$, it follows that $\angle CBC' = 90^\circ$. So by Pythagorean Theorem, $CC'^2 = BC^2 + BC'^2$, thus $AC^2 = BC^2 + CD^2$.

3 Solutions to Chapter 3 Examples

Problem 3.1 In the diagram below, points E, B, C all lie on a unit circle with center O.

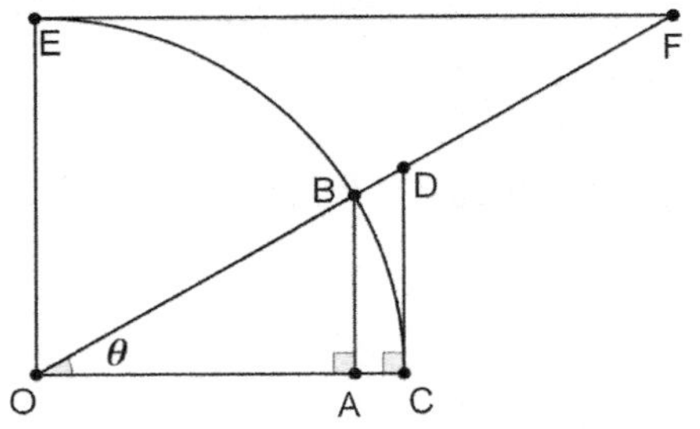

Calculate the following in terms of θ:

1. AO
2. AB
3. CD
4. DO
5. EF

Answer

Using SOH-CAH-TOA we have:

1. $\cos(\theta)$
2. $\sin(\theta)$
3. $\tan(\theta)$
4. $\sec(\theta)$
5. $\cot(\theta)$

Problem 3.2 Write the following in terms of $\sin(\theta), \cos(\theta)$, etc.

1. $\sin(90° - \theta)$
2. $\cos(90° - \theta)$
3. $\sin(180° + \theta)$
4. $\cos(180° + \theta)$
5. $\sin(180° - \theta)$
6. $\cos(180° - \theta)$
7. $\sin(-\theta)$

8. $\cos(-\theta)$

Answer

Using the unit circle definitions:

1. $\cos(\theta)$
2. $\sin(\theta)$
3. $-\sin(\theta)$
4. $-\cos(\theta)$
5. $\sin(\theta)$
6. $-\cos(\theta)$
7. $-\sin(\theta)$
8. $\cos(\theta)$

Problem 3.3 Prove the Law of Sines.

Solution

Start by drawing an altitude $\overline{CD}$ from C, so $\triangle ADC$ and $\triangle BDC$ are both right triangles. We thus have $CD = b\sin(A)$ and also $CD = a\sin(B)$. Rearranging gives $\dfrac{a}{\sin(A)} = \dfrac{b}{\sin(B)}$. Indentical arguments hold with altitudes from A and B, which yield the other equalities.

Problem 3.4 Let ABC be an equilateral triangle, and let P be a point in the interior of the triangle. Given that $PA = 3$, $PB = 4$, and $PC = 5$, find the side length of $\triangle ABC$.

Answer

$\sqrt{25+12\sqrt{3}}$.

Solution

Rotate $\triangle APB$ by 60 degrees so that the side $\overline{AB}$ coincides with side $\overline{AC}$, and the new triangle is $AP'C$, where P' is the point that P moves to. Now $\triangle APP'$ is equilateral, so $PP' = 3$. We also have $P'C = 4$, and $PC = 5$, thus $\angle PP'C = 90°$. So $\angle AP'C = 150°$. Using Law of Cosines, $AC = \sqrt{3^2+4^2-2\cdot 3\cdot 4\cos 150°} = \sqrt{25+12\sqrt{3}}$.

Problem 3.5 In triangle ABC, $AB = c$, $BC = a$, and $CA = b$. Suppose that $(a+b+c)(a+b-c) = 3ab$. Determine the measure of $\angle C$.

Answer

60°.

Solution

$(a+b)^2-c^2=3ab$, so $c^2=a^2+b^2-ab$. According to Law of Cosine, $\cos\angle C=\frac{1}{2}$, therefore $\angle C=60^\circ$.

Problem 3.6 A triangle has sides $\sqrt{3}-1<x<\sqrt{6}$ and contains angles of 15° and 45°. Use this information to calculate $\sin(15^\circ)$.

Answer

$\frac{\sqrt{3}-1}{2\sqrt{2}}$.

Solution

The last angle is 120°. The smallest and largest sides are opposite the smallest and largest angles, so using the Law of Sines we have

$$\frac{\sin(15^\circ)}{\sqrt{3}-1}=\frac{\sin(120^\circ}{\sqrt{6}}.$$

Since $\sin(120^\circ)=\frac{\sqrt{3}}{2}$ we solve for $\sin(15^\circ)=\frac{\sqrt{3}-1}{2\sqrt{2}}$.

Problem 3.7 Points A,B,C,D lie in the plane. Suppose C lies between A,B on $\overline{AB}$ while D is not on line $\overleftrightarrow{AB}$. If $CD=4, BC=4, AC=5$, and $BD=6$, what is AD?

Answer

6.

Solution

Use the Law of Cosines to get $4^2=4^4+6^2-2\cdot4\cdot6\cos(\angle CBD)$ and solve to get $\cos(\angle CBD)=\frac{3}{4}$. Hence using the Law of Cosines again we have $(AD)^2=6^2+9^2-2\cdot 6\cdot9\cdot\cos(\angle CBD)=36+81-108\cdot\frac{3}{4}=36$ so $AD=6$.

Problem 3.8 Solve the following equations. Express your answers in radians.

(a) $2\tan(x) = \sec(x)$.

Answer

$\pm\dfrac{\pi}{6} + 2\pi k$ for all integers k.

Solution

Multiplying both sides by $\cos(x)$ we get $2\sin(x) = 1$, so $\sin(x) = 1/2$.

(b) $2\sin\left(2x - \frac{\pi}{2}\right) = \sqrt{2}$

Answer

$\pm\dfrac{3\pi}{8} + \pi k$ for all integers k.

Solution

Let $z = 2x$. Note $\sin(z - \pi/2) = -\sin(\pi/2 - z) = -\cos(z)$. Solving $\cos(z) = -\sqrt{2}/2$ we have $z = \pm 3\pi/4 + 2\pi k$ for all integers k. Hence $x = \pm 3\pi/8 + \pi k$ for all integers k.

(c) $\sin(x) + \sin(5x) - 2 = 0$.

Answer

$\dfrac{\pi}{2} + 2\pi k$ for all integers k.

Solution

Note that replacing x by $x + 2\pi$ results in the same equation, so we can focus on solutions between 0 and 2π radians. Since $\sin(x), \sin(5x) \le 1$, we must have $\sin(x) = \sin(5x) = 1$. $\sin(x) = 1$ implies that $x = \pi/2$, while $\sin(5x) = 1$ implies that $x = \pi/10, \pi/2, 9\pi/10, 13\pi/10, 17\pi/10$. The only common solution is $\pi/2$, so our final solution is $x = \pi/2 + 2\pi k$ for integers k.

Problem 3.9 Suppose $ABCD$ is a square with side length 2. Suppose a line containing A intersects side CD at E and the line extending BC at F. Find (with proof) the value of $\dfrac{1}{AE^2} + \dfrac{1}{AF^2}$.

Answer

$\frac{1}{4}$.

Solution

First note that since AD is parallel to BC, we have $\angle DAE = \angle BFA$, call this angle θ. We have $\frac{AD}{AE} = \frac{2}{AE} = \cos(\theta)$ and $\frac{AB}{AF} = \frac{2}{AF} = \sin(\theta)$. Hence

$$\frac{1}{AE^2} + \frac{1}{AF^2} = \frac{\cos^2(\theta) + \sin^2(\theta)}{2^2} = \frac{1}{4},$$

as needed.

Problem 3.10 Suppose you have a convex quadrilateral with diagonals c, d which intersect at an angle of θ. Find a formula for the area of the parallelogram.

Answer

$\frac{1}{2}cd\sin(\theta)$.

Solution

Suppose the diagonals divide c into c_1, c_2 and d into d_1, d_2 as shown below.

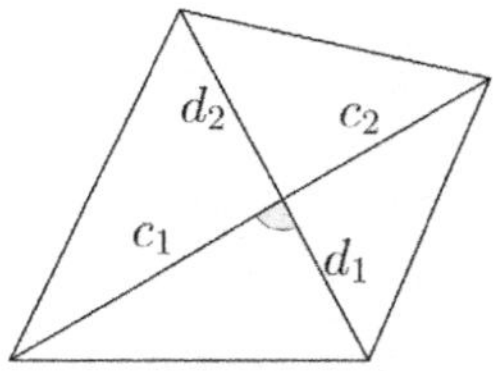

Therefore, the quadrilateral is made up of four triangles, with respective areas

$$\frac{1}{2}c_1d_1\sin(\theta), \frac{1}{2}c_1d_2\sin(180^\circ - \theta), \frac{1}{2}c_2d_2\sin(\theta), \frac{1}{2}c_2d_1\sin(180^\circ - \theta).$$

As $\sin(\theta) = \sin(180^\circ - \theta)$ and $c = c_1 + c_2, d = d_1 + d_2$ the sum of these four triangles is $\frac{1}{2}cd\sin(\theta)$.

4 Solutions to Chapter 4 Examples

Problem 4.1 Prove that $\sin(x+y) = \sin(x)\cos(y) + \cos(x)\sin(y)$. The diagram below may be helpful:

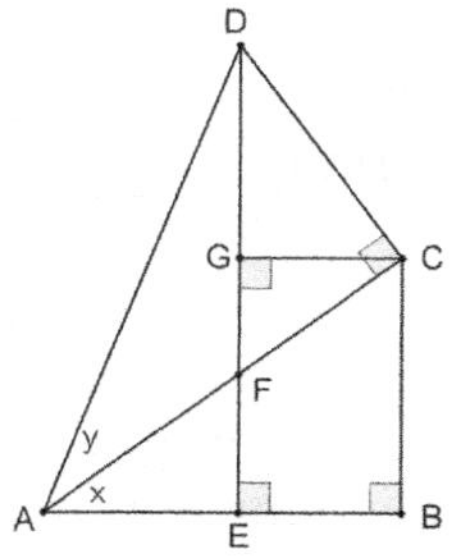

Hint: $DE = DG + GE$.

Solution

We have $DE = DG + GE = DG + BC$. We have $BC = AC \cdot \sin(x)$ and $DG = CD \cdot \sin(90^\circ - x) = CD \cdot \cos(x)$ (as $\angle GCD = 90^\circ - \angle GCF$ and $\angle GCF = x$). Therefore

$$\begin{aligned}\sin(x+y) &= \frac{DE}{AD} = \frac{AC \cdot \sin(x) + CD \cdot \cos(x)}{AD} \\ &= \sin(x) \cdot \frac{AC}{AD} + \cos(x) \cdot \frac{CD}{AD} = \sin(x)\cos(y) + \cos(x)\sin(y),\end{aligned}$$

by examining the diagram.

Problem 4.2 Prove the following. (Hint: Use previous formulas!)

(a) The sine difference formula.

Solution

Follows from the Sum and Even-Odd formulas.

(b) The sine double angle formula.

Solution

Follows from the Sum formula.

(c) The cosine double angle formulas.

Solution

Follows from the Sum formula.

Problem 4.3 Prove the following.

(a) The tangent sum and difference formulas.

Solution

Follows from the definition of tangent (sine over cosine) and the Sum/Difference formulas after dividing the numerator and denominator by $\cos(x)\cos(y)$.

(b) The tangent double angle formulas.

Solution

Follows from the Sum formula.

Problem 4.4 Compute the following values.

(a) $\cos(225^\circ)$.

Answer

$\frac{-\sqrt{2}}{2}$.

Solution

$\cos(225^\circ) = \cos(360^\circ - 225^\circ) = \cos(135^\circ) = -\cos(180^\circ - 135^\circ) = -\cos(45^\circ)$.

(b) $\sin(75^\circ)$.

Answer

$\frac{1+\sqrt{3}}{2\sqrt{2}} = \frac{\sqrt{2}+\sqrt{6}}{4}$.

Solution

Use either the fact that $75 = 45 + 30$ and the sum formula or the fact that $75 = 150/2$ and the half angle formula.

(c) $\cos(165°)$.

Answer

$$-\frac{1+\sqrt{3}}{2\sqrt{2}} = -\frac{\sqrt{2}+\sqrt{6}}{4}.$$

Solution

We have $\cos(165°) = \sin(90° - 165°) = \sin(-75°) = -\sin(75°)$.

(d) $\sec(525°)$.

Answer

$\sqrt{2} - \sqrt{6}$.

Solution

Note $\cos(525°) = -\cos(15°)$. Then use the half angle formula.

Problem 4.5 Simplify $\cos(A+B)\cos(B) + \sin(A+B)\sin(B)$.

Answer

$\cos(A)$.

Solution

Note that using the cosine difference formula, the above is equal to $\cos((A+B) - B)$.

Problem 4.6 In $\triangle ABC$, suppose $AB = 15$, $CD = 13$ and $\cos(B) = \frac{33}{65}$. What is $\sin(C)$?

Answer

$\frac{12}{13}$.

Solution

Using the Law of Cosines we calculate

$$AC^2 = 13^2 + 15^2 - 2 \cdot 13 \cdot 15 \cdot \frac{33}{65} = 196,$$

so $AC = 14$. Using the Law of Sines we have $\frac{\sin(C)}{15} = \frac{\sin(B)}{14}$ so $\sin(C) = \frac{15\sin(B)}{14}$.
Lastly,

$$\sin(B)^2 = 1 - \frac{33^2}{65^2} = \frac{65^2 - 33^2}{65^2} = \frac{(65-33)(65+33)}{65^2} = \frac{56^2}{65^2},$$

so $\sin(C) = \frac{15}{14} \cdot \frac{56}{65} = \frac{12}{13}$.

Problem 4.7 Trig. Identities Practice

(a) Show that $\sin(2x) + \sin(2y) = 2\sin(x+y)\cos(x-y)$.

Solution

Starting with the right hand side, we use the sum and difference formulas to write it as

$$2(\sin(x)\cos(y) + \cos(x)\sin(y)(\cos(x)\cos(y) + \sin(x)\sin(y)).$$

Multiplying out we have

$$2(\sin(x)\cos(x)\cos^2(y) + \sin^2(x)\sin(y)\cos(y)$$
$$+\cos^2(x)\sin(y)\cos(y) + \sin(x)\cos(x)\sin^2(y)).$$

Factoring we can simplify this to $2\sin(x)\cos(x) + 2\sin(y)\cos(y) = \sin(2x) + \cos(2x)$ as needed.

(b) Given $\triangle ABC$ (sides a, b, c), prove $b\cos(B) + c\cos(C) = a\cos(B-C)$.

Solution

Similar to an earlier problem, the Law of Sines we have that $a = k\sin(A), b = k\sin(B), c = k\sin(C)$ for a constant k. Hence (using the double angle formula and the identity from part (a))

$$b\cos(B) + c\cos(C) = k\sin(B)\cos(B) + k\sin(B)\cos(B) = \frac{k}{2}(\sin(2B) + \sin(2C)$$
$$= k\sin(B+C)\cos(B-C)$$

Note that $\sin(B+C) = \sin(180^\circ - A) = \sin(A)$ so we have $k\sin(A)\cos(B-C) = a\cos(B-C)$ as needed.

Problem 4.8 Find all solutions x (in radians) so that $0 \le x < 2\pi$ to the following equations.

(a) $\sin(2x) = \tan(x)$.

Answer

$x = 0, \pi/4, 3\pi/4, \pi, 5\pi/4, 7\pi/4$.

Solution

Using the double angle formula we can rewrite everything in terms of $\sin(x), \cos(x)$:

$$2\sin(x)\cos(x) = \frac{\sin(x)}{\cos(x)}.$$

Hence either $\sin(x) = 0$, so $x = 0, \pi$, or $\cos^2(x) = \pm\frac{1}{2}$ so $\cos(x) = \frac{\sqrt{2}}{2}$. This gives $x = \pi/4, 3\pi/4, 5\pi/4, 7\pi/4$ for a total of 6 solutions in $0 \le x < 2\pi$.

(b) $\cos(2x) = \cos(x)$.

Answer

$x = 0, 2\pi/3, 4\pi/3$.

Solution

The double angle formula gives $2\cos^2(x) - 1 = \cos(x)$ or $2\cos^2(x) - \cos(x) - 1(2\cos(x) + 1)(\cos(x) - 1) = -$ so $\cos(x) = -1/2$ or $\cos(x) = 1$. Thus $x = 2\pi/3, 4\pi/3$ or $x = 0$.

(c) $\sin(x) + \cos(x) = \frac{\sqrt{6}}{2}$. Hint: $\cos(\pi/4) = \sin(\pi/4) = \sqrt{2}/2$.

Answer

$$x = \frac{\pi}{12}, \frac{5\pi}{12}.$$

Solution

Using the hint we can multiply both sides by $\sqrt{2}/2$ to get

$$\sin(x)\cos(\pi/4) + \cos(x)\sin(\pi/4) = \frac{\sqrt{3}}{2}.$$

Note then the left hand side is $\sin(x+\pi/4)$. Thus, $x+\pi/4 = \pi/3, 2\pi/3$ so $x = \pi/12, 5\pi/12$.

Problem 4.9 Suppose in triangle ABC we have $\angle A = 45^\circ$, $AB = 3$ and $AC = 2\sqrt{2}$. What is $\tan(B)$?

Answer

2.

Solution

Using the Law of Cosines we have $BC^2 = 9+8-12\sqrt{2}\cos(45^\circ) = 5$ so $BC = \sqrt{5}$. Similarly, using the Law of Cosines again we can calculate $\cos(B) = \dfrac{\sqrt{5}}{5}$. Since $\sin^2(B)+\cos^2(B) = 1$ we get $\sin(B) = \dfrac{2\sqrt{5}}{5}$. Hence $\tan(B) = \dfrac{\sin(B)}{\cos(B)} = 2$.

Problem 4.10 Show that in a triangle ABC (with sides a,b,c) we have $\dfrac{a-b}{a+b} = \tan\left(\dfrac{A-B}{2}\right)\tan\left(\dfrac{C}{2}\right)$. Hints: Start by writing the left hand side in terms of sines. An identity we proved earlier might be useful again here!

Solution

By the Law of Sines we have that $a = k\sin(A), b = k\sin(B), c = k\sin(C)$ for a constant k (we will see exactly what this constant is in the coming weeks). Hence,

$$\frac{a-b}{a+b} = \frac{k\sin(A)-k\sin(B)}{k\sin(A)+k\sin(B)} = \frac{2\sin\left(\frac{A-B}{2}\right)\cos\left(\frac{A+B}{2}\right)}{2\sin\left(\frac{A+B}{2}\right)\cos\left(\frac{A-B}{2}\right)}$$

5 Solutions to Chapter 5 Examples

Problem 5.1 Using only the basics about parallel lines and congruent/similar triangles and the fact that the area of a rectangle is bh, prove the following. (Note: once a fact is proven below, you can use it in later parts.)

(a) The area of a parallelogram is bh.

Solution

As in the diagram below, cutting a triangle at one end of the parallelogram and moving it to the other results in a $b \times h$ rectangle.

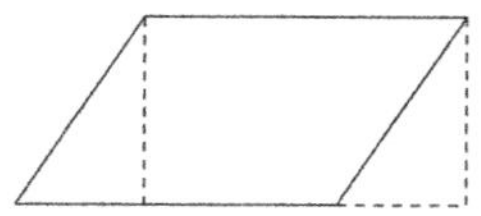

(b) The area of a triangle is $\frac{1}{2}bh$ (prove this two ways!).

Solution

Method 1: A right triangle is half a rectangle, so has area $\frac{1}{2}bh$. Then any triangle can be split into two right triangles (by dropping an altitude to the longest side).
Method 2: Two copies of any triangle can be combined to form a parallelogram with base b and height h.

(c) The area of a trapezoid is $\frac{b_1+b_2}{2}h$.

Solution

Draw a diagonal of the trapezoid. This breaks the trapezoid into two triangles, each with height h and having bases b_1 and b_2.

(d) The area of a trapezoid is also mh where m is the *median* of the trapezoid, which connects the midpoints of the two non-parallel sides.

Solution

As in the diagram below, cutting triangles below the median and rotating them above the median results in a $m \times h$ rectangle.

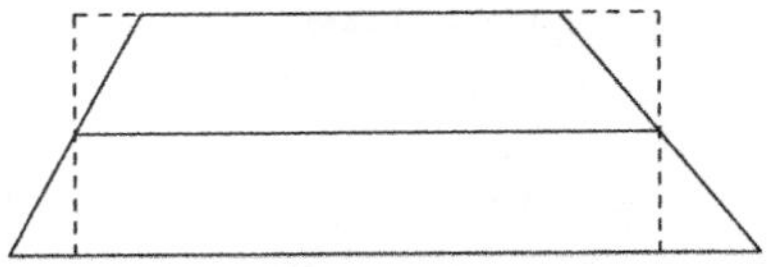

Problem 5.2 What is the ratio of areas between two similar triangles? Prove your result!

Solution

The square of the ratio between the corresponding sides.
Note that heights of similar triangles are also in ratio of the corresponding sides (as the altitude divides the similar triangles into smaller similar triangles).

Problem 5.3 Prove the Pythagorean Theorem using areas.

Solution

There are literally hundreds of different proofs. The simplest: Let $\triangle ABC$ be a right triangle where $\angle C$ is a right angle. Label the sides with a, b, c in the standard way. Draw the altitude $\overline{AD}$ on the hypotenuse. Then the three triangles, $\triangle ABC$, $\triangle ADC$ and $\triangle BDC$ are all similar. Their side ratio is $a : b : c$ based on the hypotenuses. Thus their area ratio is $a^2 : b^2 : c^2$. Since the two smaller triangle's areas add up to that of the big triangle, we have $a^2 + b^2 = c^2$.

Problem 5.4 In $\triangle ABC$, $AB > AC > BC$, $\overline{CD}, \overline{BE}, \overline{AF}$ are altitudes on $\overline{AB}, \overline{AC}, \overline{BC}$, respectively. Show that $CD < BE < AF$.

Solution

The area of $\triangle ABC$ can be calculated in three ways: $[ABC] = \frac{1}{2}AB \cdot CD = \frac{1}{2}AC \cdot BE = \frac{1}{2}BC \cdot AF$, so $CD/BE = AC/AB$ and $AF/BE = AC/BC$, thus gives the proof.

Problem 5.5 In triangle ABC, $AC = 10$, $BC = 24$, $AB = 26$. What is the altitude on $\overline{AB}$?

Answer

$\frac{120}{13}$.

Solution

Note $\triangle ABC$ is a right triangle. Therefore the area is $10 \cdot 24/2 = 120$. Since $AB = 26$ the area is also $26h/2 = 13h$ where h is the altitude on $\overline{AB}$, we have $h = 120/13$.

Problem 5.6 Let $ABCD$ be a parallelogram, and E, H, F, G be points on sides $\overline{AB}, \overline{BC}, \overline{CD}, \overline{DA}$ respectively, and $\overline{EF} \| \overline{BC}$ and $\overline{GH} \| \overline{AB}$. Let P be the intersection of $\overline{EF}$ and $\overline{GH}$. If $[GPFD] = 10, [PHCF] = 8, [EBHP] = 16$, find $[ABCD]$.

Answer

54.

Solution

Note $\overline{EF}, \overline{GH}$ divide $ABCD$ into four parallelograms, whose areas are proportional. That is $[AEPG]/[GPFD] = [EBHP]/[PHCF]$ so $[AEPG] = 10 \cdot 16/8 = 20$. Hence the total area of $ABCD$ is 54.

Problem 5.7 Suppose you only know that the centroid exists. Prove (using areas!) that the centroid divides each median in a ratio of $1:2$.

Solution

Consider the following diagram with the medians intersecting at G:

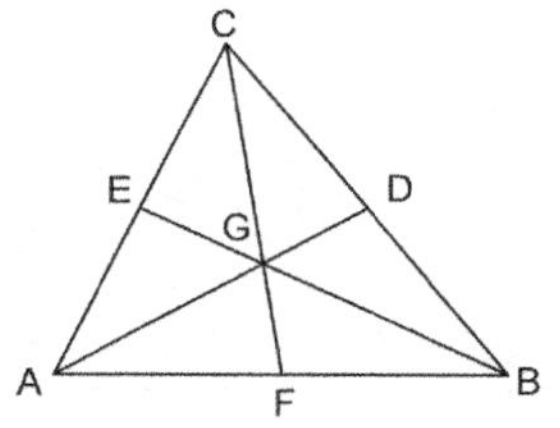

Note first that $[ACF] = [BCF]$ and $[AGF] = [BGF]$ (F is a midpoint and they share respective heights). Hence $[ACG] = [ACF] - [AGF] = [BCF] - [BGF] = [BCG]$. An identical argument gives $[ACG] = [ABG]$. Hence, $[AFG] = [ABG]/2 = [ACG]/2$ so in particular $[AFG] = [AFC]/3$. These two triangles share a height from A, so their bases FG, FC must be in ratio $1:3$ and hence $FG : GC = 1:2$. The other medians are handled identically.

Problem 5.8 Let $ABCD$ be a parallelogram, with midpoints E, F, G, H (say on $\overline{AB}, \overline{BC}, \overline{CD}, \overline{DA}$). Let I, J be the midpoints of $\overline{EF}, \overline{GH}$. Find the area of $\triangle JIG$ as a fraction of the area of $ABCD$.

Answer

$1/8$.

Solution

We will use the facts (which you should be able to prove!) $\overline{FH} \| \overline{DC}$ and $\overline{EF} \| \overline{GH}$. We have $[JIG] = [HIG]/2$ (same height, J midpoint), $[HIG] = [HFG]$ ($\overline{EF} \| \overline{GH}$), $[HFG] = [DCFH]/2$ ($\overline{FH} \| \overline{DC}$), and finally $[DCFH] = [ABCD]/2$ (F, H midpoints). Combining these give $[JIG] = [ABCD]/8$.

Problem 5.9 Prove that if a triangle has side lengths a, b, c, inradius r, and circumradius R we have $2Rr = \dfrac{abc}{a+b+c}$.

Solution

We have that the area of the triangle is $abc/4R$ and sr where s is the semiperimeter, so $abc = 4srR$. Rearranging gives the desired result.

Problem 5.10 Let $ABCD$ be a parallelogram as in the diagram, with E the midpoint of $\overline{BC}$.

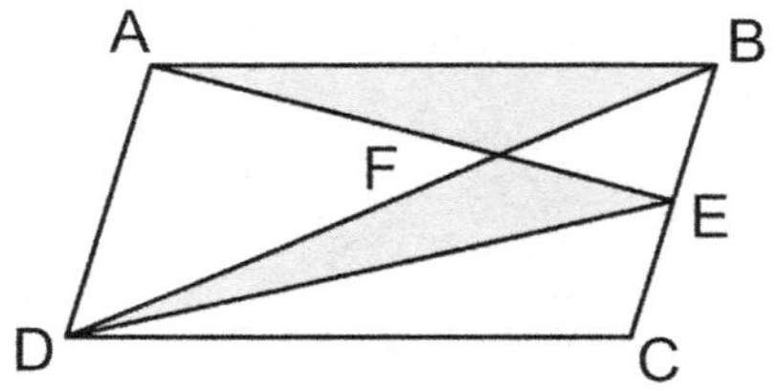

(a) Compare the shaded regions $\triangle ABF$ and $\triangle DEF$, which one has the larger area?

Answer

The have the same area.

Solution

Note $[ABD] = [DEF]$ as they have the same height and base. Hence, after removing the shared $\triangle AFD$ we see the two shaded triangles have the same area.

(b) Find the area of the shaded regions $\triangle ABF$ and $\triangle DEF$ in terms of the entire area of the parallelogram.

Answer

$[ABCD]/3$.

Solution

First note (since E is a midpoint) that $[DEC] = [BED] = [ABCD]/4$. Since $ABCD$ is a parallelogram and $BE = AD/2$, $\triangle AFD \sim \triangle EFB$, with ratio of side lengths $2:1$. Using this information, we have $[DFE] = 2[BEF]$. Therefore, $[DFE] = [ABF] = [ABCD]/6$. Hence the sum of the shaded regions is $[ABCD]/3$.

6 Solutions to Chapter 6 Examples

Problem 6.1 Ceva's Theorem and Converse.

(a) Ceva's Theorem and Its Converse

Solution

Let G be the intersecting point of the three cevians. Then

$$\frac{BX}{XC} = \frac{[ABX]}{[AXC]} = \frac{[GBX]}{[GXC]} = \frac{[ABX]-[GBX]}{[AXC]-[GXC]} = \frac{[ABG]}{[ACG]},$$

Similarly $\frac{CY}{YA} = \frac{[BCG]}{[BAG]}$, and $\frac{AZ}{ZB} = \frac{[CAG]}{[CBG]}$. Then the product is 1.

(b) Prove the converse to Ceva's Theorem using an indirect proof.

Solution

Let G be the intersection of AX and BY. Connect and extend CG to intersect with size AB at point Z'. Then the cevians AX, BY, and CZ' are concurrent. Based on Ceva's Theorem, $\frac{BX}{XC} \cdot \frac{CY}{YA} \cdot \frac{AZ'}{Z'B} = 1$. This means $\frac{AZ}{ZB} = \frac{AZ'}{Z'B}$. Thus Z and Z' are the same point.

Problem 6.2 (Centers of Triangles) Prove the following, using Ceva's Theorem (or its converse) if applicable.

(a) (Centroid) In every triangle ABC, the three medians (i.e. the line from a vertex to the midpoint of the opposite side) are concurrent. This point is called the *centroid* of the triangle ABC.

Solution

Trivial by the converse of Ceva's Theorem.

(b) (Incenter) In every triangle ABC, the three angle bisectors (i.e. the line from a vertex bisecting the angle at that vertex) are concurrent. This point is called the *incenter* of the triangle ABC, because it is the center of the circle inscribed in triangle ABC.

Solution

Use Angle Bisector Theorem on each angle bisector, then the converse of Ceva's Theorem.

Problem 6.3 Suppose you have a trapezoid $ABCD$ with $\overline{AB}$ parallel to $\overline{CD}$. Let E be the intersection of the diagonals. Suppose $AB = 10, CD = 15$ and $\triangle ADE$ has area 24. Find the area of $ABCD$.

Answer

100.

Solution

Since $\overline{AB} \| \overline{CD}$, $\triangle ABE \sim \triangle CDE$, with ratio of corresponding sides $10 : 15 = 2 : 3$. Hence $DE : EB = 3 : 2$ and since $\triangle AED, \triangle AEB$ share the same height from A, $[AED] : [AEB] = 3 : 2$ so $[AEB] = 24 \cdot \frac{2}{3} = 16$. We can use a similar argument to get $[DEC] = 36$ and $[BEC] = 24$. Hence the total area is 100.

Problem 6.4 Suppose the altitudes of a triangle are in ratio $2 : 2 : 3$ and the triangle has a perimeter of 24. Find the area of the triangle.

Answer

$18\sqrt{2}$.

Solution

As in an earlier problem, we use the fact that we can calculate the area using all of the altitudes. Suppose the altitudes (from A, B, C respectively) are h_A, h_B, h_C with $h_A : h_B : h_C = 2 : 2 : 3$. If a, b, c are the opposite sides, we then have (after cancelling $1/2$): $a \cdot h_A = b \cdot h_B = c \cdot h_C$. We therefore have that $b : a = h_A : h_B = 2 : 2$ and $c : b = h_B : h_C = 3 : 2$. Hence, the sides are in ratio $c : b : a = 2 : 3 : 3$. If the perimeter of the triangle is 24, this means the sides of the triangles are $6, 9, 9$. Using Heron's formula, the area is thus, $18\sqrt{2}$.

Problem 6.5 Let G be the centroid of $\triangle ABC$, and $AG = 3, BG = 4, CG = 5$, find $[ABC]$.

Answer

18.

Solution

Say the medians are $\overline{AD}, \overline{BE}, \overline{CF}$, which divide $\triangle ABC$ into six equal area triangles. Extend $\overline{BE}$ 2 units further, giving a point H. Note $AGCH$ is a parallelogram, so $CH = 3$. Hence, $\triangle GCH$ is a right triangle with area 6. Note that $[GCH] = [ACG] = [ABC]/3$ and thus $[ABC] = 18$.

Problem 6.6 (2002 AMC 12A #22) Triangle ABC is a right triangle with $\angle ACB$ as its right angle, $m\angle ABC = 60°$, and $AB = 10$. Let P be randomly chosen inside $\triangle ABC$, and extend $\overline{BP}$ to meet $\overline{AC}$ at D. What is the probability that $BD > 5\sqrt{2}$?

Answer

$\frac{3-\sqrt{3}}{3}$.

Solution

Triangle ABC is a 30-60-90 triangle, and $BC = 5$, $AC = 5\sqrt{3}$. Let E be the point on $\overline{AC}$ such that $CE = 5$. Then $BE = 5\sqrt{2}$. The desired probability is for P to fall in the triangle ABE, and equals the ratio $\frac{ABE}{ABC} = \frac{AE}{AC} = \frac{5\sqrt{3}-5}{5\sqrt{3}} = \frac{3-\sqrt{3}}{3}$.

Problem 6.7 Suppose a dodecagon is in a square as in the diagram below (the vertices of the dodecagon on the square are the midpoints of the sides):

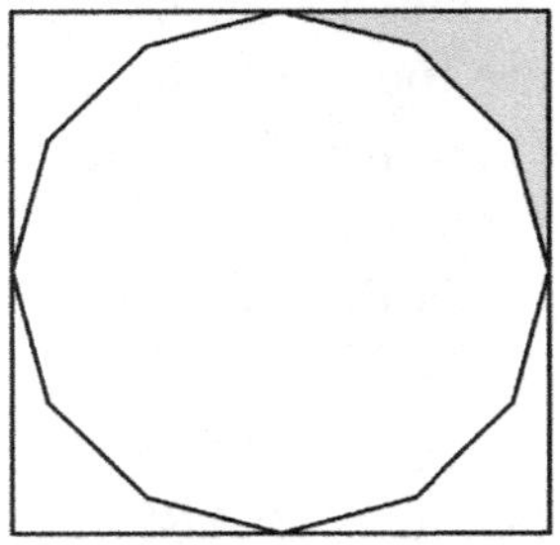

If the area of the square is 4, what is the area of the shaded region?

Answer

$\frac{1}{4}$.

Solution

Consider the following diagram

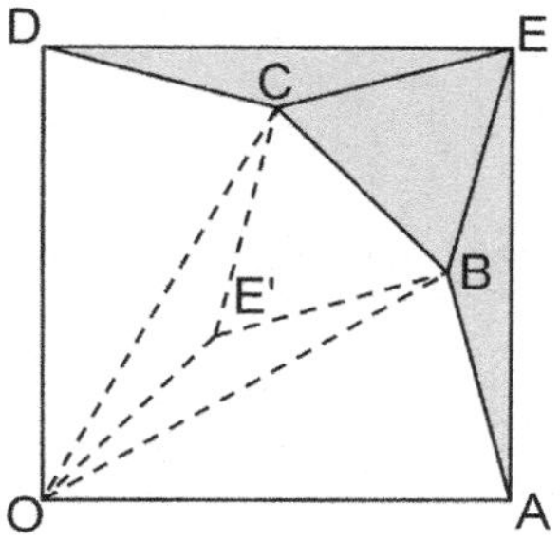

where $OABC$ is a quarter of the dodecagon and E' is E reflected across $\overline{BC}$. $\triangle OAB, \triangle BOC, \triangle COD$ are all congruent isosceles triangles with angles $30°, 75°, 75°$. Clearly $\triangle BCE \cong \triangle BCE'$ and calculating angles we see $\triangle ABE, \triangle ECD, \triangle OE'B, \triangle CE'O$ are all congruent isosceles triangles with angles $15°, 15°, 150°$. Hence the shaded region has area $[OAED]/4$ which is $1/16$th of the entire square. Hence the shaded region has area $1/4$.

Problem 6.8 Let $ABCDE$ be a convex pentagon. Suppose further that the triangle cut off by each diagonal has area 1. What is the area of the full pentagon $ABCDE$?

Answer

$\frac{5+\sqrt{5}}{2}$.

Solution

We are given that $[ABC] = [BCD] = [CDE] = [DEA] = [EAB] = 1$. Note $\triangle EAB$ and $\triangle ABC$ share the base AB so they have the same heights. This implies that $\overline{AB}$ is parallel to $\overline{CE}$. A similar argument gives $\overline{BC} \parallel \overline{AD}$, $\overline{CD} \parallel \overline{BE}$, etc. Thus, we get the following type of diagram.

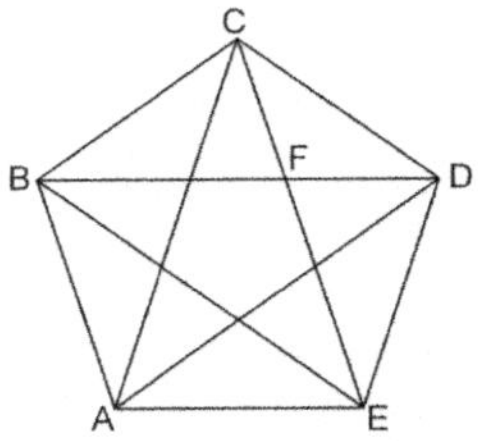

Set $[BCF] = x$. Then $[CFD] = 1 - x$. Using these triangles we have

$$\frac{BF}{FD} = \frac{[BCF]}{[CFD]} = \frac{x}{1-x}.$$

Also, note $[BFE] = [ABE] = 1$ as $ABFE$ is a parallelogram. Hence using triangles $\triangle BFE$ and $\triangle DFE$,

$$\frac{BF}{FD} = \frac{[BFE]}{[DFE]} = \frac{1}{x}.$$

Hence $\dfrac{x}{1-x} = \dfrac{1}{x}$ so we can solve for $x = (\sqrt{5}-1)/2$ (ignoring the negative root). Lastly,

$$[ABCDE] = [ABE] + [BFE] + [BCF] + [CDE] = 1 + 1 + x + 1 = 3 + x = \frac{5+\sqrt{5}}{2}.$$

Problem 6.9 Start with trapezoid $ABCD$. Extend $\overline{AD}, \overline{BC}$ to meet at O as in the diagram below.

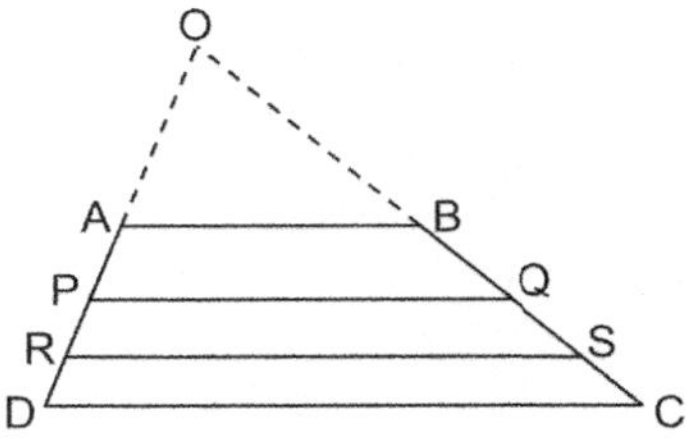

Suppose $AD = 1, OA = 1$. Construct P, Q, R, S such that $\overline{AB} \parallel \overline{PQ} \parallel \overline{RS}$ and $[ABQP] = [QSRP] = [SCDR]$. What are AP, PR, RD?

Answer

$AP = \sqrt{2} - 1, PR = \sqrt{3} - \sqrt{2}, RD = 2 - \sqrt{3}.$

Solution

Note that $\triangle OAB \sim \triangle OPQ \sim \triangle ORS \sim \triangle ODC$. Since $OA = AD$ we have $[OAB] = [ABCD]/4$. Hence we want $[OPQ] = [ABCD]/2 = 2[OAB], [ORS] = 3[ABCD]/4 = 3[OAB]$. Thus, $OP = \sqrt{2}, OR = \sqrt{3}$ and hence $AR = \sqrt{2}-1$, $PR = \sqrt{3}-\sqrt{2}$, $RD = 2-\sqrt{3}$ as needed.

Problem 6.10 (Kurrah's Theorem) Let $\triangle ABC$ be given. Construct D, E as in the diagram below with $\angle ABC = \angle ADB = \angle BEC$.

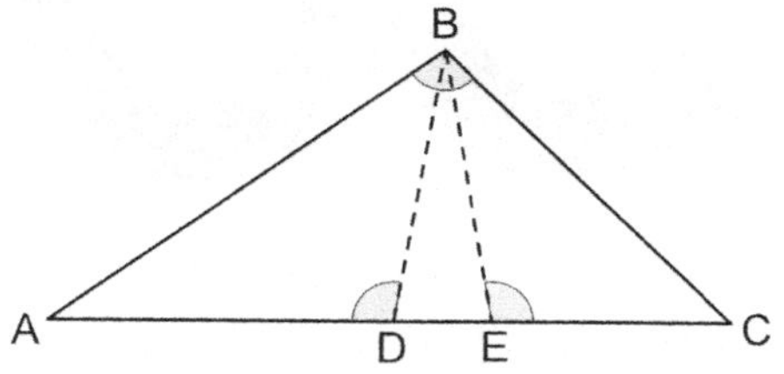

Prove that

$$AB^2 + BC^2 = AC(AD + CE).$$

Solution

Note we have $\triangle ABC \sim \triangle ADB \sim \triangle BEC$. Thus

$$\frac{AB}{AC} = \frac{AD}{AB} \Rightarrow AB^2 = AC \cdot AD$$

and

$$\frac{BC}{AC} = \frac{CE}{BC} \Rightarrow BC^2 = AC \cdot CE.$$

Adding we get $AB^2 + BC^2 = AC(AD + CE)$ as needed.

7 Solutions to Chapter 7 Examples

Problem 7.1 Storing Tires

(a) Suppose two tires, each with radius 1ft rest upright on the ground and touching each other, as pictured below:

How much space is needed horizontally to store the tires?

Answer

4.

Solution

Note the distance between the center of the circles is parallel to the ground, so the total distance needed is the two diameters, which sum to 4ft.

(b) Repeat part (a) with two tires of radius $1,2$ feet respectively.

Answer

$3+2\sqrt{2}$.

Solution

To store the tires using the least amount of horizontal space, we again examine the line $\overline{AB}$ connecting the two centers, as in the diagram below:

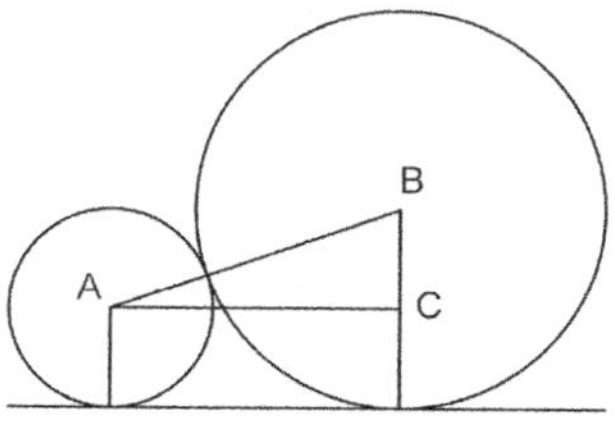

From here we see the horizontal distance in between the centers of the tires is AC, where $\triangle ABC$ is a right triangle with $AB = 1+2 = 3, BC = 2-1 = 1$. Hence $AC = 2\sqrt{2}$ and the total distance needed is $1+2\sqrt{2}+2$.

Problem 7.2 Suppose you start with a circle of radius 1. Draw another circle of radius 1 with center an arbitrary point on the first circle. Let R denote the region consisting of all points that are inside both circles.

(a) Find the perimeter of R.

Answer

$4\pi/3$.

Solution

Let A, B be the centers of the circle, and C, D be the intersection points of the two circles. Hence the perimeter of R consists of the arc from C to D (containing A) plus the arc from D to C (containing B). Further, as both $\triangle ABC, \triangle ABD$ are equilateral triangles (SSS), each arc is seen to be $120°$, or $1/3$ of a circle. Hence the perimeter of R is $2 \cdot \frac{1}{3} \cdot 2\pi = \frac{4\pi}{3}$.

(b) Find the area of R.

Answer

$2\pi/3 - \sqrt{3}/2$.

Solution

Consider the labeling from above. We then have that the area of R is the sum of the areas of the two sectors minus the two equilateral triangles: $2 \cdot \frac{1}{3} \cdot \pi - 2 \cdot \frac{\sqrt{3}}{4}$.

Problem 7.3 Inscribing Circles in Sectors

(a) What is the radius of the largest circle that can fit in a quarter circle of radius 1?

Answer

$\sqrt{2}-1$.

Solution

Consider the following general diagram:

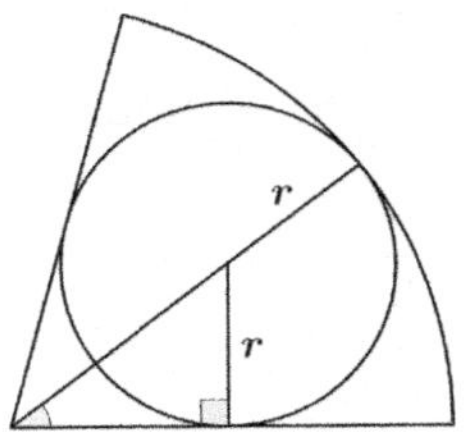

where if the sector is θ°, the marked angle in the right triangle is $\theta/2$.
In this case, $\theta = 90^\circ$, so the triangle is a 45-45-90 triangle with hypotenuse $r\sqrt{2}$. Hence, $1 = r + r\sqrt{2}$ so solving for r gives $r = \dfrac{1}{1+\sqrt{2}} = \sqrt{2}-1$.

(b) What if instead you are fitting it into a 60°-sector?

Answer

$1/3$.

Solution

A similar method to part (a) gives the equation $1 = r + 2r$ so $r = 1/3$.

Problem 7.4 Let $\angle APB$ be an inscribed angle on a circle with center O. Prove that $\angle APB$ is half the angular size of arc $\widehat{AB}$ if:

(a) O lies on $\angle APB$.

Solution

Assume O lies on $\overline{AP}$. Then $\triangle OPB$ is isosceles, so $2\angle BPO = 2\angle BPA = \angle BOA$ as needed.

(b) O lies inside $\angle APB$.

Solution

Let Q be such that $\overline{PQ}$ is a diameter. Note $\triangle BOP$ is isosceles, so $\angle BOQ = 2\angle BPO$. Similarly, $\angle AOQ = 2\angle APO$. Hence, $\angle BPA = \angle BPO + \angle APO = (\angle BOQ + \angle AOQ)/2 = \angle BOA/2$ as needed.

Problem 7.5 Prove that if two chords AC, BD intersect inside a circle at point P then the measure of $\angle APB$ is half the sum of the angular sizes of $\widehat{AB}, \widehat{CD}$.

Solution

Draw $\overline{BC}$. We have $\angle APB = \angle DBC + \angle ACB$ equals have the sum of the angular sizes of $\widehat{CD}, \widehat{AB}$ (as $\angle DBC, \angle ACB$ are inscribed angles).

Problem 7.6 Suppose $\widehat{AB}$ is an arc with angular size 60° and CD is a diameter such that if rays $\overrightarrow{BA}, \overrightarrow{DC}$ are extended to intersect at a point E, $\angle AEC = 30$. Find the angular size of arc $\widehat{BD}$.

Answer

90°.

Solution

Let the angular measure of $\widehat{BD} = x$. Then the size of $\widehat{AC} = 180 - 60 - x = 120 - x$. Then $\angle AEC = 30 = \dfrac{x - (120 - x)}{2}$ so solving for x gives $x = 90^\circ$.

Problem 7.7 Suppose two perpendicular chords intersect and divide each other in a ratio of 1 : 2. Find the radius of the circle if each chord is 12in long.

Answer

$2\sqrt{10}$.

Solution

The two chords divide each other in segments of 4 and 8 inches. The perpendicular bisector of each chord goes through the center of the circle, so we can form a right triangle with sides 2 and 6 with hypotenuse r, the radius of the circle. Hence, $r = \sqrt{2^2+6^2} = 2\sqrt{10}$.

Problem 7.8 Suppose ω is a circle with radius 6 and center O. Let $\widehat{AB} = 135^\circ$. Let C be on ω such that $\overline{OA} \parallel \overline{BC}$. Find $[OACB]$.

Answer

$18+9\sqrt{2}$.

Solution

Let E on $\overline{BC}$ be such that $\overline{OE}$ is a height. As $\overline{OE} \perp \overline{BC}$ and O is the center of the circle, E is the midpoint of $\overline{BC}$. Further, $\triangle OEB$ is an isosceles right triangle. Hence, the trapezoid has bases $6, 6\sqrt{2}$ and height $3\sqrt{2}$, so it has area $\frac{1}{2} \cdot (6+6\sqrt{2}) \cdot 3\sqrt{2} = 18+9\sqrt{2}$.

Problem 7.9 (AMC 12A 2007 #10) A triangle with side length in the ratio $3:4:5$ is inscribed in a circle of radius 3. What is the area of the triangle?

Answer

$\frac{216}{25}$.

Solution

Let the sides of the triangle have lengths $3x$, $4x$, and $5x$. The triangle is a right triangle, so its hypotenuse is a diameter of the circle. Thus $5x = 2 \cdot 3 = 6$, so $x = 6/5$. The area of the triangle is $\frac{1}{2} \cdot 3x \cdot 4x = \frac{216}{25}$.

Problem 7.10 Let $\mathscr{C}_1$ and $\mathscr{C}_2$ be circles defined by

$$(x-10)^2 + y^2 = 36$$

and

$$(x+15)^2 + y^2 = 81,$$

respectively. What is the length of the shortest line segment $\overline{PQ}$ that is tangent to $\mathscr{C}_1$ at P and to $\mathscr{C}_2$ at Q?

Answer

20

Solution

The circle $\mathscr{C}_1$'s center is $(10,0)$ and radius 6, and $\mathscr{C}_2$ center $(-15,0)$ and radius 9. Calculate the length of the internal tangent line. This line passes through the origin (why?). There are two triangles of the 3-4-5 side-ratios with actual side lengths $(6,8,10)$ and $(9,12,15)$, and the length of the tangent line is $8+12=20$. The external tangent line is longer: $\sqrt{25^2-(9-6)^2}=\sqrt{616}$.

8 Solutions to Chapter 8 Examples

Problem 8.1 Prove that if $\angle A + \angle C = \angle B + \angle D = 180^\circ$ in quadrilateral $ABCD$, then $ABCD$ is a cyclic quadrilateral .

Solution

Start by circumscribing a circle about $\triangle ABC$. If D is not on the circle, then $\angle B + \angle D$ is either greater or less than 180° (depending on whether D is inside or outside the circle.

Problem 8.2 Suppose $ABCD$ is a cyclic quadrilateral.

(a) Prove that $\angle ABD = \angle ACD$ (and similarly $\angle BAC = \angle BDC$, etc.).

Solution

$\angle ABD, \angle ACD$ both share the arc $\widehat{AD}$ (not containing B, C), so they are the same angle.

(b) Prove that if E is the intersection of $\overline{AC}, \overline{BD}$ then $AE \cdot EC = BE \cdot ED$.

Solution

Using part (a), we have that $\triangle ABE \sim \triangle CED$, so $AE/DE = BE/EC$ and the result follows.

Problem 8.3 Power of a Point Inside a Circle

(a) Suppose $\overline{YZ}$ is a chord, with X the midpoint. Prove the Power of a Point formula for X and the line through Y, Z.

Solution

Recall we want to show $XY \cdot XZ = XO^2 - R^2$. Since the lengths are directed, we need to show $R^2 = XO^2 + YX \cdot XZ$ which is clear from the Pythagorean Theorem and because X is the midpoint.

(b) Prove the Power of a Point formula for arbitrary X inside the circle.

Solution

Suppose X and a line that intersects the circle at Y, Z are given. Choose Y', Z' on the circle such that X is the midpoint of $X'Y'$. By part (a), $XY' \cdot XZ' = XO^2 - R^2$. However, Y, Z, Y', Z' are concyclic, so $XY' \cdot XZ' = XY \cdot XZ$ (note we do not need to worry about directed lengths, as both quantities are negative). Combining these two gives the desired result.

Problem 8.4 Prove Ptolemy's theorem, using the following diagram as guidance,

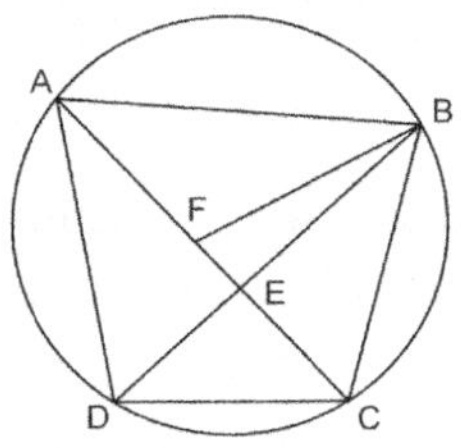

where F is such that $\angle ABF = \angle CBD$.

Solution

We constructed F such that $\triangle ABF \sim \triangle DBC$, and we also have $\triangle ABD \sim FBC$ (make sure you can fill in the details here). Therefore, we have that $AF \cdot BD = AB \cdot CD, CF \cdot BD = BC \cdot DA$. Adding these equalities and simplifying gives $AC \cdot BD = AB \cdot CD + AD \cdot BC$. as needed.

Problem 8.5 Suppose chords $\overline{AB}, \overline{CD}$ intersect at E, such that $AE : EB = 1 : 3$ and $CE : ED = 1 : 12$. Find the ratio of $AB : CD$.

Answer

$8/13$.

Solution

Let $AE = x, CE = y$, so $EB = 3x, ED = 12y$. By Power of a Point, $3x^2 = 12y^2$, so solving for x/y we get $x/y = 2$. Hence, $\dfrac{AB}{CD} = \dfrac{4x}{13y} = \dfrac{4}{13} \cdot 2 = \dfrac{8}{13}$.

Problem 8.6 (AHSME 1999 #21) A circle is circumscribed about a triangle with sides 20, 21, and 29, thus dividing the interior of the circle into four regions. Let A, B, and

C denote the areas of the non-triangular regions, with C being the largest. Compute $C-(A+B)$.

Answer

210.

Solution

Since 20, 21, and 29 form a Pythagorean triple, the triangle is a right triangle, and the side with length 29 is the diameter. It is easy to see that the result of $C-(A+B)$ is exactly the area of $\triangle ABC$.

Problem 8.7 Let A,B,C,D be four points, arranged in clockwise order, on circle ω. Segments AC and BD intersect at P. Given that $AB=3, BP=4, PA=5, PC=6$, find the radius of circle ω.

Answer

$\sqrt{565}/4$.

Solution

We find the diameter. From the given lengths it is easy to see that $\triangle ABP$ is a right triangle, and $\angle ABP=90^\circ$. So AD is the diameter. By Power of a Point, $AP\cdot PC=BP\cdot PD$, so $PD=5\cdot 6/4=15/2$. In right triangle ABD, apply the Pythagorean theorem to get $AD=\sqrt{AB^2+AD^2}=\sqrt{565}/2$. Thus the radius is $\sqrt{565}/4$.

Problem 8.8 Suppose we have a rectangle $ABCD$ with $AB=8$, $BC=12$. Inscribe a circle in the rectangle so that it touches sides $\overline{AB},\overline{BC},\overline{AD}$. Let M be the midpoint of $\overline{AB}$. Call $E\neq M$ the intersection of $\overline{MD}$ with the circle. Find DE.

Answer

$8\sqrt{10}/5$.

Solution

Let F be such that $\overline{AD}$ is tangent to the circle at F. Since $AB=8$, the radius of the circle is 4. Hence, $AM=4$ and $DF=12-4=8$. By Power of a Point, $DF^2=DM\cdot DE$. As $DM=\sqrt{4^2+12^2}=4\sqrt{10}$, $DE=\dfrac{8^2}{4\sqrt{10}}=\dfrac{8\sqrt{10}}{5}$.

Problem 8.9 Given right triangle ABC, construct semicircles on the three sides as shown.

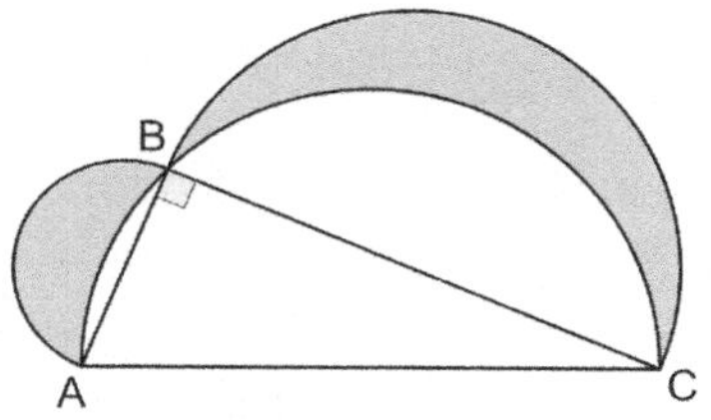

Given that $AB = 5, BC = 12$. Find the sum of the areas of the shaded regions.

Answer

30

Solution

The sum of the areas of the two smaller semicircles is $\frac{1}{2}\left(\frac{\pi}{4}(AB^2+BC^2)\right) = \frac{\pi}{8}(AB^2+BC^2) = \frac{\pi}{8}AC^2$, which is exactly the area of the large semicircle. Subtracting the common regions from both sides, the remaining regions are the shaded regions on the side of two smaller semicircles, and the right triangle ABC on the large semicircle. So the sum of the areas of the shaded regions equals the area of the triangle The area $[ABC] = \frac{1}{2}\cdot 5\cdot 12 = 30$. So the answer is 30.

Problem 8.10 In $\triangle ABC$, $AB = 37, AC = 58$. Use A as center and AB as radius, draw a circle to intersect $\overline{BC}$ at D where D is between B and C.

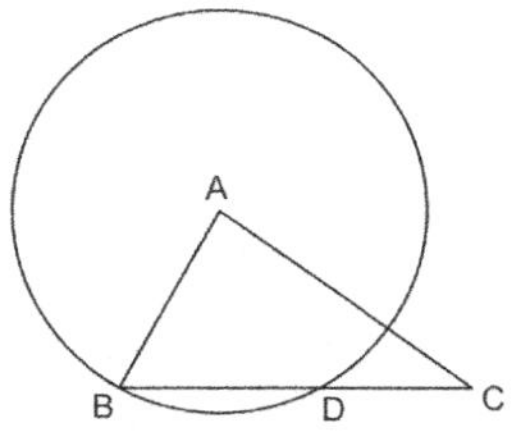

Given that the lengths of $\overline{BD}$ and $\overline{DC}$ are both integers, compute BC.

Answer

57

Solution

Let E be the intersection of $\overline{AC}$ and the circle. Extend $\overline{CA}$ to intersect the circle at F. So $AE = AF = 37$, $CF = 58 + 37 = 95$, and $CE = 58 - 37 = 21$. By Power of a Point, $CB \cdot CD = CF \cdot CE = 95 \times 21 = 1995$. Since the lengths of $\overline{BD}$ and $\overline{DC}$ are integers, so is the length of $\overline{BC}$. Since $\overline{CE}$ goes through A, the center of the circle, $CF > CB$. Since $1995 = 3 \times 5 \times 7 \times 19$, the only way that 1995 is expressed as the product of two factors, and both factors are less than 95, is $1995 = 57 \times 35$. Thus $BC = 57$ and $CD = 35$. Hence 57 is the final answer.

9 Solutions to Chapter 9 Examples

Problem 9.1 Inscribing Spheres in a Cone

(a) Find the volume of the largest sphere that can fit inside a cone of radius 1 and height $\sqrt{3}$.

Answer

$\dfrac{4\sqrt{3}}{27}\pi.$

Solution

Using a 30-60-90 triangle we the radius of the sphere is equal to $1/\sqrt{3}$.

(b) Assume the sphere as in (a) is placed inside the cone. Suppose you now want to fit another sphere in the cone that is tangent to the base. Find the radius of the largest such sphere.

Answer

$\dfrac{\sqrt{3}}{9}.$

Solution

Let $s = 1/\sqrt{3}$ be the radius in (a). Let s' be the radius of the new smaller sphere. Using a 30-60-90 triangle again we have $2(s-s') = s+s'$, so $3s' = s$ and hence $s' = 1/(3\sqrt{3})$.

Problem 9.2 Suppose you have a regular tetrahedron with side length a.

(a) Show that the surface area is $a^2\sqrt{3}$.

Solution

A regular tetrahedron is made up of 4 equilateral triangles. Each of these has area $a^2\sqrt{3}/4$ so the total surface area is $a^2\sqrt{3}$.

(b) Show that the volume is $\dfrac{a^3\sqrt{2}}{12}$.

Solution

Note the result follows if we show a regular tetrahedron (which is a triangular pyramid) has height $a\sqrt{2}/\sqrt{3}$. Recall the apex is above the incenter of the base. For an equilateral triangle, the incenter is $a/\sqrt{3}$ from the vertex. This leads to the following diagram:

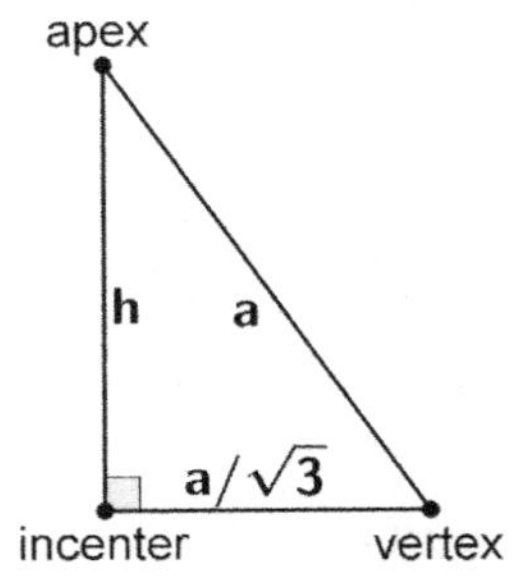

Hence, solving for h, we get $h = a\sqrt{2/3}$ as needed.

Problem 9.3 (2010 AMC 10A #20) Suppose a bored bee lives on a cube with side length 1. For "fun" he decides to visit every vertex of the cube, each exactly once, starting and ending at the same vertex. It will travel from one vertex to another using straight lines (either crawling or flying). Give an example of a path that uses the maximum distance and find this distance.

Answer

$4\sqrt{2}+4\sqrt{3}$.

Solution

The distance between opposite corners of the cube is $\sqrt{3}$ and the distance opposite corners of a square face is $\sqrt{2}$. Note any path has 8 stops, and it is possible to travel only using the above distances, 4 times each. This is optimal, as there are only 4 pairs of opposite corners (and the bee can only use each once). An example path is given below:

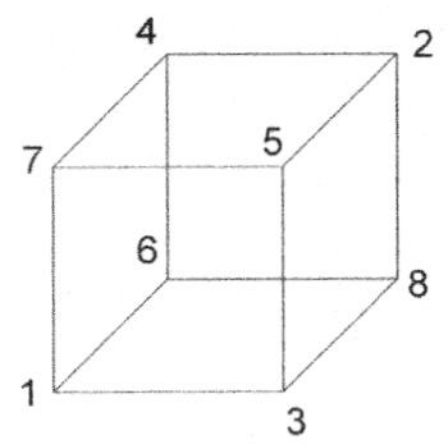

(The bee travels from 1 to 2 to 3 ... to 8 to 1.)

Problem 9.4 Four identical balls (spheres), each of radius 1in, are glued to the ground so that their centers form the vertices of a square with side length 2in. Suppose you rest a fifth identical ball on the four balls (so the fifth ball is a sphere externally tangent to the other spheres). How far does this ball rest off the ground?

Answer

$\sqrt{2}$.

Solution

Using symmetry we look at spheres with centers across the diagonal:

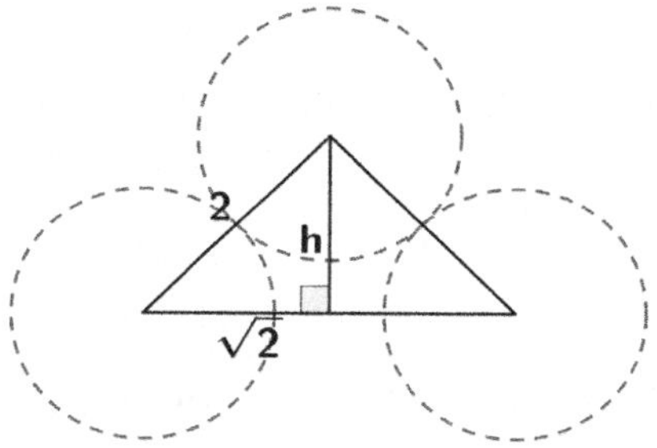

Let $h+1$ denote the height of the center of the fifth sphere off the ground. Setting up right triangles gives us $2^2 = 2 + h^2$, so solving for h gives $\sqrt{2}$.

Problem 9.5 Suppose you pick 4 vertices of a cube to form a tetrahedron.

(a) How many different (non-congruent) tetrahedra are possible? Are any of them regular tetrahedra?

Answer

4

Solution

First assume, 3 of the vertices are contained in the same square face. Label the cube $ABCD - A'B'C'D'$ so that the vertices chosen are A, B, C, and one of A', B', C', D'. Note $ABC - B' \cong ABC - C'$, so this leads to 3 different tetrahedron, none of which are regular.

The only other possibility (up to congruence) is (again using the labelling $ABCD - A'B'C'D'$) $AC - B'D'$. Note this tetrahedron *is* regular.

(b) Find the volumes for each of the possibilities in (a) if the cube has volume 1.

Answer

$1/6$ or $1/3$

Solution

If 3 vertices are contained in the same square face, then we have a tetrahedron (which is a triangular pyramid) with base of area $1/2$ and height of 1. Thus, the volume is $\frac{1}{3} \cdot \frac{1}{2} \cdot 1 = \frac{1}{6}$.

Now suppose we have tetrahedron $AC - B'D'$. Note this tetrahedron divides the rest of the cube into 4 congruent tetrahedra, which we just saw have volume $\frac{1}{6}$. Hence, tetrahedron $AC - B'D'$ has volume $\frac{1}{3}$.

Problem 9.6 Suppose you have a regular square pyramid $S - ABCD$ with height 6 whose square has side length 4. Call the midpoints of the square E, F, G, H (on $\overline{AB}, \overline{BC}, \overline{CD}, \overline{DA}$ respectively) and the midpoints of $\overline{SA}, \overline{SB}, \overline{SC}, \overline{SD}$ respectively T, U, V, W. Form polyhedron $EFGH - TUVW$ (with 10 faces).

(a) Describe the faces of $EFGH - TUVW$. How many vertices and edges does the polyhedron have?

Solution

There are two square faces (both different), and 8 triangular faces (2 groups of 4 congruent triangles). There are 8 vertices and 16 edges.

(b) Find the volume of $EFGH - TUVW$. Hint: Do this indirectly.

Answer

20.

Solution

Note the pyramid $S-ABCD$ is made up of (i) the polyhedron $EFGH-TUVW$, (ii) the regular square pyramid $S-TUVW$, and (iii) four congruent triangular pyramids (for example $T-AEH$. The volume of the entire pyramid is $\frac{1}{3}\cdot 4^2\cdot 6=32$. The volume of the smaller square pyramid is $\frac{1}{3}\cdot 2^2\cdot 3=4$ (using similar triangles). Lastly, the volume of each triangular pyramid is $\frac{1}{3}\cdot\frac{1}{2}\cdot 2^2\cdot 3=2$. Hence, the volume we want is $32-4-4\cdot 2=20$.

Problem 9.7 Suppose you have a unit cube. Pick two opposite corners. In each corner, form a tetrahedron using the corner and the three adjacent vertices. Remove these two tetrahedra and call the resulting polyhedron $\mathscr{S}$.

(a) How many vertices, edges, and faces does the resulting polyhedron have? Describe the faces.

Answer

6 vertices, 12 edges, and 8 faces.

Solution

Note we are removing 2 vertices, not changing the number of edges, and adding 2 faces. All of the faces are triangles.

(b) Find the volume of $\mathscr{S}$.

Answer

$2/3$.

Solution

Note each of the tetrahedra removed have base area $1/2$ and height 1, hence has volume $1/6$. Hence $\mathscr{S}$ has volume $1-2\cdot\frac{1}{6}=\frac{2}{3}$.

Problem 9.8 Suppose you have tetrahedron $S-ABC$. Cut the tetrahedron with a plane between S and $\triangle ABC$, forming a new tetrahedron $S-A'B'C'$ (with A' on $\overline{SA}$, etc.). Prove that the ratio of volumes of $S-ABC$ to $S-A'B'C'$ is equal to $SA \cdot SB \cdot SC : SA' \cdot SB' \cdot SC'$.

Solution

View the tetrahedra as having bases $\triangle SAB$ and $\triangle SA'B'$. Let h, h' be the corresponding heights. Using similar triangles we have $h : h' = SC : SC'$. Further, since triangles $\triangle SAB, \triangle SA'B'$ share an angle, we have $[SAB] : [SA'B'] = SA \cdot SB : SA' \cdot SB'$. As the volume of a tetrahedron is the base area times the height, the result follows.

Problem 9.9 (2012 AMC 10A #21) Let points $A=(0,0,0)$, $B=(3,0,0)$, $C=(0,4,0)$, $D=(0,0,5)$. Suppose E,F,G,H are midpoints of (respectively) $\overline{BD}, \overline{AB}, \overline{AC}, \overline{DC}$. Prove that $EFGH$ is a square.

Solution

Form tetrahedron $D-ABC$ and note that $\overline{DA}$ is perpendicular to the plane containing $\triangle ABC$ (the xy-plane). Using similar triangles, E,H are directly above F,G (that is, $\overline{EF}, \overline{HG}$ are perpendicular to the xy-plane) and $EF = AD/2 = GH$. If follows that $EFGH$ is a rectangle. Then note $EF = GH = \frac{AD}{2} = \frac{5}{2}$ and $FG = EH = BC/2 = \frac{\sqrt{3^2+4^2}}{2} = \frac{5}{2}$.

Problem 9.10 Suppose you start with a right cone and cut off the top of the cone with a plane parallel to the base. The resulting solid, called a *frustrum* has two circular "bases", say with radii R and r (with $R > r$), and height h. (Hence from the side the frustrum looks like a trapezoid with bases R, r and height H.)

(a) Show that the volume of a frustrum is $\frac{\pi H}{3}(R^2 + Rr + r^2)$.

Solution

Let $H+h$ denote the height of the original cone. Clearly the volume of the frustrum is $\pi R^2(H+h)/3 - \pi r^2 h/3$. Using similar triangles we have $\frac{h}{r} = \frac{h+H}{R}$ so $h = \frac{rH}{R-r}$. Simplifying gives the above result.

(b) Suppose a sphere can be inscribed in a frustrum with base radii r, R such that the sphere is tangent to the two bases and the side. Find the radius of such a sphere in terms of r, R.

Answer

$\sqrt{rR}$.

Solution

Let s denote the radius of the sphere. We have the following diagram:

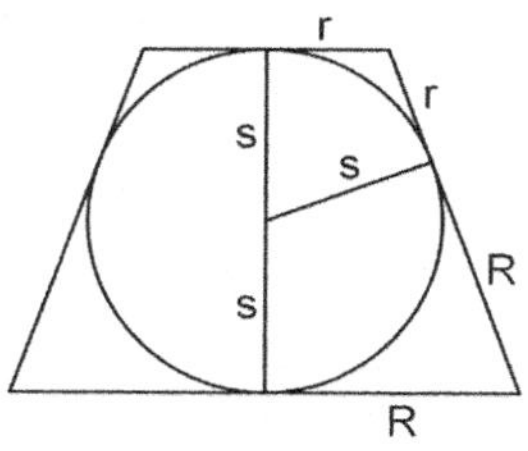

Therefore, $H = 2s$. Further, dropping a height from an endpoint of the top base yields a right triangle, so $(R+r)^2 = (2s)^2 + (R-r)^2$. Solving for s gives $s = \sqrt{rR}$.

Made in the USA
Monee, IL
15 April 2023

31801505R00083